W9-AAE-054

LIVES AND TIMES OF GREAT PIONEERS IN CHEMISTRY

(LAVOISIER to SANGER)

LIVES AND TIMES OF GREAT PIONEERS IN CHEMISTRY

(LAVOISIER to SANGER)

C N R Rao

Indumati Rao

Jawaharlal Nehru Centre for Advanced Scientific Research, Bangalore, India

NEW JERSEY • LONDON • SINGAPORE • BEIJING • SHANGHAI • HONG KONG • TAIPEI • CHENNAI • TOKYO

Published by

World Scientific Publishing Co. Pte. Ltd.
5 Toh Tuck Link, Singapore 596224
USA office: 27 Warren Street, Suite 401-402, Hackensack, NJ 07601
UK office: 57 Shelton Street, Covent Garden, London WC2H 9HE

British Library Cataloguing-in-Publication Data
A catalogue record for this book is available from the British Library.

LIVES AND TIMES OF GREAT PIONEERS IN CHEMISTRY
(Lavoisier to Sanger)

ISBN 978-981-4689-05-2
ISBN 978-981-4689-92-2 (pbk)

Printed in Singapore by Mainland Press Pte Ltd.

Writing this book has been a labour of love.

We dedicate the book to

all those who appreciate, admire and enjoy chemistry.

- Authors

Preface

Chemistry got to be recognized as a bonafide subject only towards the end of the 18th century – thanks to Lavoisier. It then made great progress in the traditional branches of chemistry such as organic, inorganic and analytical chemistry. Physical chemistry got duly absorbed as part of the main stream in the early part of the 20th century. 1920s and 30s saw a major change in the nature of chemistry because of the impact of structure and bonding. The chemical bond became a center piece of discussion in chemistry. Many new physical techniques became available for the study of structure. Structure, dynamics and synthesis were considered to be the three main components of chemistry in the 1960-70 period, and advances in chemistry pertained to a great extent to structure and to the discovery of new reactions as well as their mechanisms. Since the early 1970s, a significant change has occurred wherein biology and advanced materials have become major components of chemistry. Today, chemistry has gone beyond the molecular frontier and areas such as self-organization and assembly, complex systems, atmosphere, synthetic biology, and high-level computation have become areas of serious exploration. It was our desire to pay tribute to the great pioneers of chemistry by putting together a book which gives adequate coverage of the times and lives of these chemists besides their contributions.

In this work, we have written about some of the great chemical pioneers whose ideas and contributions have helped to create chemistry as we know it. We start with Lavoisier (Father of chemistry) and end with Sanger. After Lavoisier in the 18th century, we had Dalton and Faraday in the 19th century. Faraday is considered by many to be the greatest scientist of all time. In the early part of the 20th century, chemists created waves with important discoveries related to chemical bonding, molecular structure, reactivity and dynamics, besides synthesis. We have written extensively about chemists in this golden era. In the early part of the 20th century, Mendeleev proposed the periodic table. In organic chemistry, we had Willstätter, Robinson and Woodward who are considered to be the trinity in organic chemistry. Two of the great 20th century chemists who originated many new ideas are G. N. Lewis and Linus Pauling. The 21 chemists we have written about belong to different nationalities: Britain-6, France-1,

Germany-5, Netherlands-1, Russia-1, Sweden-2, Switzerland-1 and United States-4.

As we went through the wonderful journey tracing the growth of chemistry and the lives of the fine people who contributed to it, we have noted the unique attributes, personal idiosyncrasies and philosophical attitudes of these creators. One cannot help noticing the romantic relationship between chemistry and Lavoisier, the devotional approach of Faraday, a great teacher's qualities in Ostwald, an imperious attitude in Arrhenius, an architect's zeal in Lewis, the crusading spirit of a conqueror in Pauling and so on. The romantic involvement of a poet one sees in Lavoisier seems to slowly give way to an accomplishment-oriented attitude of craftsmen and artists as time has gone on. It is noteworthy that most of the pioneers were born in ordinary (even poor) families and became great through their zeal and hard work. Many of them had to overcome serious obstacles to reach the high level of attainment. They had social commitment as well.

It is possible that we have omitted a few pioneers who in the judgment of others should have been included in this book. We would like to be excused for such omission. We ourselves can think of a few chemists who could have been included but we decided to limit the number to 21. We do hope that the readers will pardon us for possible omissions and errors in judgment, and read the book as a recollection of the glorious past of chemistry and an expression of gratitude to the great pioneers. In order to make the book interesting and "real", we have cited extensively from various sources. We have given due credit to the original sources and we earnestly request to be forgiven in case of any mistakes or omissions in citing the literature.

We sincerely thank the invaluable contribution of Jatinder Kaur in formatting and in preparing the print-ready manuscript. Thanks are also due to Sanjay S. R. Rao for his assistance.

Bangalore (India)
2015

C N R Rao
Indumati Rao

CONTENTS

1. ANTOINE LAVOISIER (1743–1794)

Father of chemistry

The year 1994 marked the 200th death anniversary of Lavoisier. In the preface to the article *The life and legacy of Antoine Laurent Lavoisier* to commemorate the life and legacy of the father of modern chemistry, Peter Childs wrote "*It is more than 200 years this year the greatest chemist of all time - Antoine-Laurent Lavoisier – became a victim of the irrational mob fury of the French Revolution. Lavoisier died at the age of 51 at the height of his creative powers and after establishing modern chemistry on a firm footing. Who knows what more he might have achieved had he lived? Every modern chemist, indeed every chemist for the last 200 years, lives in the shadow and on the legacy of Lavoisier. It is instructive 200 years later to look back at Lavoisier's life and achievements, to the very birth of modern chemistry. Before he died the phlogiston theory was dead....and the oxygen theory of combustion and respiration was firmly established. Lavoisier laid his personal claim to be the author of the chemical revolution in 1792*" (From Childs [1]).

(From *Understanding Chemistry*)

Chemistry was not a well-defined subject in the early 18th century and Lavoisier propelled chemistry and established it as an experimental science.

"The coping-stones of eighteenth century chemistry, which are at the same time the foundation-stones of the modern science, were laid by Antoine Laurent Lavoisier (1743-1794)....." (From Holmyard [2]).

Lavoisier is believed to have said *"I am young and avid for glory"* (From the Website scienceworld.wolfram.com [3]) when he was a young student. He lived up to this self-belief in every role he played during his lifetime. Lavoisier was a polymath in the renaissance mould of Leonardo da Vinci.

Early years

Antoine Laurent Lavoisier was born on August 26, 1743 in an affluent family. His father, Jean Antoine Lavoisier, was a wealthy and successful lawyer and was close to the powerful aristocracy as he was also an *avocatau parlement* (parliamentary counsel). His mother Emille Punctis was the daughter of a rich attorney. Lavoisier lost his mother when he was five years old and went to live with his maternal grandmother and aunt. His aunt devoted her life to him and upon her death, left her considerable wealth to him. The double inheritance made him independently wealthy and came as a boon when the time came to choose a career. He could choose any career that he fancied without worrying about the payslip. Lavoisier's father was particular that young Lavoisier should get the best possible education that combined the rigors of science with liberal education of humanities that inculcated social awareness. He enrolled young Lavoisier in College des Quatre Nations when he was 11 years old as a day scholar, to receive a holistic and liberal education there. Later at the College Mazarin under a number of inspiring teachers Lavoisier studied (in addition to mathematics, astronomy) botany (de Jussieu), geology (Guettard), chemistry (Rouelle), languages and literature. He not only gained theoretical knowledge in science subjects by attending their lectures but also gained practical skills by working in laboratories and doing field work with Guettard. This unique blend of proficiency in science and classical and liberal arts contributed to Lavoisier's breadth of interest in science and commitment to social justice that he displayed throughout his professional life. He also graduated in law in 1763 and obtained a license to practice law in 1764. He was expected to follow the family tradition and take up law as his profession, but his path to glory was elsewhere.

Lavoisier grew up during the period of French enlightenment and was inspired by its ideals. His was a rare intellect that could combine the demands of public office with the pursuit of science. He was attracted to the excitement of science as well as to the need to work for social causes. He studied geology for four years from 1763 to1767 under Guettard who was a member of the Royal Academy of Sciences. During this period, he toured the Alsace-Lorraine region and conducted experiments on the geology of the rocks of the region. Young Lavoisier was fascinated by everything he saw around him and took copious notes on the vegetation, nature of soils, daily temperatures and mines. On the basis of this study, he presented a paper on the physical and chemical properties of hydrated calcium sulfate (gypsum) to the French Academy of Sciences in 1764. His interest in the problems facing the general public in Paris at that time led him to study the problem of inadequate street lighting which led to social problems. In 1766, he submitted an essay about these problems and possible solutions to the competition organized by the French Academy of Sciences. It is said that he studied various types of lamps and compared the different kinds of oils and wicks used by them. He drew detailed diagrams to show the amount of light generated and the area lit up by different lamps. He also presented in his essay the cost of maintenance. Though he did not win a prize, the judges were so impressed by the thoroughness of the presentation and the original solutions offered by Lavoisier that they gave him a special gold medal which was presented by King Louis XV. The Academy took the usual step of publishing the essay. In recognition of these contributions, he was elected to the French Academy of Sciences in 1768 when he was only 25 years old.

Around this time, he turned his attention to chemistry as he was fascinated by the well known French chemist Pierre-Joseph Macquer's *Dictionnaire de chymie* published in 1766. However, it was the prominent French scholar, Etienne Condillac, who inspired his life-long passion and devotion to chemistry.

When he became 26 years old, it became imperative for Lavoisier to choose a career. Rather than take up a job with the French Academy, Lavoisier bought stock in *Ferme Generale*, a private company that collected agricultural taxes in lieu of the advance (based on the estimated profit made by the farming community) made to the king and he became a tax collector. While this choice of career afforded him the time and finances to pursue chemistry, it proved to be a fatal choice 26 years later!

In 1775, Lavoisier was appointed to Gun Powder Commission as one of the four commissioners. Their brief was to devise methods of improving the quality of the gun powder and also to improve its supply to the government. This appointment entitled him to a large house at the Royal Arsenal where he lived from 1775 to 1792. He spent his own money to convert part of the house to one of the best equipped laboratories in France where he pursued his scientific work without any external distractions. In a short time, it became the meeting place for distinguished scientists and free thinkers from all over Europe.

Lavoisier and du Pont: Eleuthère Irénée du Pont (1771-1834), founder of the du Pont Company, was the son of Pierre du Pont a reform minded government official. When Lavoisier took up the task of reforming the Royal Gunpowder and Saltpeter Administration, he hired Pierre's son Eleuthère Irénée du Pont to work in the gunpowder factory at Essonne. Years later during the French Revolution, the du Pont family were forced to emigrate to America. They settled down in Wilmington, Delaware, where Irénée du Pont set up the original gunpowder factory which later became one of the biggest chemical industrial houses.

Marriage: In 1771, Lavoisier married Marie-Anne Pierrette Paulze, his colleague's daughter, 14 years younger to him. Though she had no exposure to the world of science, young Marie-Anne plunged into Lavoisier's world of chemistry. As Lavoisier did not know English, she learnt English so that she could translate the research papers published in England by contemporary scientists (especially those of Priestley and Cavendish) and Richard Kirwan's *Essay on Phlogiston.* She learnt painting and the art of engraving so that she could illustrate Lavoisier's work and make engravings of the instruments used by Lavoisier. She was always present in the laboratory when Lavoisier was conducting experiments and also at the meetings of scientists who visited the laboratory. In a short time she became Lavoisier's invaluable collaborator. Her drawings and illustrations added immense value to Lavoisier's writings.

(From "Antoine Lavoisier", *Wikipedia*)

"Lavoisier changed the whole structure and outlook of chemistry, the science which more than any other touches our daily lives, and which affects them in almost every phase" (From McKie [4]).

Lavoisier's Chemistry

Lavoisier considered Nature as *"a vast chemical laboratory in which all kinds of compositions and decompositions are formed"* (From Lavoisier [5]). His approach to chemistry is best described in his own words: *"We must trust to nothing but facts: These are presented to us by Nature, and cannot deceive. We ought, in every instance, to submit our reasoning to the test of experiment, and never to search for truth but by the natural road of experiment and observation"* (From Lavoisier [5]).

"*Thoroughly convinced of these truths, I have imposed upon myself, as a law, never to advance but from what is known to what is unknown; never to form any conclusion which is not an immediate consequence necessarily flowing from observation and experiment; and always to arrange the facts, and the conclusions which are drawn from them, in such an order as shall render it most easy for beginners in the study of chemistry thoroughly to understand them*" (From Lavoisier [5]).

When Lavoisier began his foray into chemistry, the subject was little understood and was so underdeveloped that it was not even considered as a serious scientific discipline (unlike physics and mathematics which were well defined disciplines). Chemistry at that time was highly descriptive and qualitative. Lavoisier changed this and established that chemistry was essentially an experimental science.

He felt that time had come to *"recall chemistry to a more rigorous method of reasoning; to strip away the facts with which this science is enriched every day from that which reasoning and prejudices add thereto; to distinguish fact and observation from that which is systematic and hypothetical; finally, to mark the limit, so to speak, to which chemical knowledge has arrived, in order that those who follow us may set out with confidence from this point to advance the science"* (From Lavoisier [6]).

Conservation of Mass: Lavoisier was convinced that the total weight of the reactants and products would be conserved in any chemical reaction. He categorically stated that *"We may lay it down as an incontestible axiom, that, in all the operations of art and nature, nothing is created; an equal quantity of matter exists both before and after the experiment; the quality and quantity of the elements remain precisely the same; and nothing takes place beyond changes and modifications in the combination of these elements. Upon this principle the whole art of performing chemical experiments depends: We must always suppose an exact equality between the elements of the body examined and those of the products of its analysis"* (From Lavoisier [5]). ... *"As the usefulness and accuracy of chemistry depends entirely upon the determination of the weights of the ingredients and products ....., too much precision cannot be employed in this part of the subject, and for this purpose, we must be provided with good instruments"* (From Lavoisier [5]). This underscored the need for accurate measurement of the reactants before the chemical reaction took place and the resultant products. As the balances available at that time could not measure accurately, Lavoisier collaborated with Laplace and developed a balance.

Understanding combustion: Lavoisier spent a number of years conducting experiments to understand combustion. At this time, combustion was explained solely by the phlogiston theory, first proposed by Stahl of Germany. Phlogiston, like the Philosopher's Stone of alchemists, was just a vague idea or concept. According to this theory, a) all combustible substances contained phlogiston which escaped when the substance was burnt, b) the degree of the substance's combustibility was directly proportional to the quantity of phlogiston present in it. Lavoisier did not believe in the phlogiston theory of combustion. He felt that chemists had *"made of phlogiston a vague principle which is not at all rigorously defined, and which, in consequence, adapts itself to all explanations in which it is wished it shall enter; sometimes it is free fire, sometimes it is fire combined with the earthy element; sometimes it passes through the pores of vessels, sometimes they are impenetrable to it; it explains both the causticity and non-causticity, transparency and opacity, colours and absence of colours..."* (From Lavoisier [6]). Lavoisier focused his attention on developing a coherent theory of combustion in which phlogiston had no place. He had been conducting experiments to understand the true nature of combustion since 1771-72 by burning various substances like sulfur and phosphorus in closed containers and had deposited a sealed letter with the Secretary of the French Academy of Sciences giving the details of his experimental findings.

The sealed note deposited with the Secretary of the French Academy (1 Nov 1772)

"About eight days ago I discovered that sulfur in burning, far from losing weight, on the contrary, gains it; it is the same with phosphorus. This increase of weight arises from a prodigious quantity of air that is fixed during combustion and combines with the vapors. This discovery, which I have established by experiments, that I regard as decisive, has led me to think that what is observed in the combustion of sulfur and phosphorus may well take place in the case of all substances that gain in weight by combustion and calcination; and I am persuaded that the increase in weight of metallic calxes is due to the same cause... This

> *discovery seems to me one of the most interesting that has been made since Stahl and since it is difficult not to disclose something inadvertently in conversation with friends that could lead to the truth I have thought it necessary to make the present deposit to the Secretary of the Academy to await the time I make my experiments public*" (From Lavoisier [7]).

In the year 1774, two events took place which proved to be the turning point in Lavoisier's pursuit of understanding the true nature of combustion. Firstly, Priestley who was visiting France, described in detail his experiment with calx of mercury (which yielded a new gas which was insoluble in water and in which the candle burned brighter and small animals were friskier) to Lavoisiers over a dinner. This was followed by a letter from Scheele of Sweden giving an account of his experiment again with calx of mercury which yielded fire air with similar properties. (As both of them believed in the phlogiston theory, they did not realize that they had discovered a new gas. Scheele named it fire air and Priestley called it dephlogisticated air). Lavoisier immediately realized the importance of this information and decided to repeat their experiments in the reverse order by first heating mercury in a sealed metal retort with common air in it for twelve days at the end of which mercury had changed into calx of mercury. He weighed the retort and the volume of air in the retort before heating it and after the conclusion of the experiment. Lavoisier quantified the change in weight of the container. He also observed that the volume of air now was ~4/5 of its original volume. Lavoisier found that when a lighted candle and small animals were introduced into the retort, the flame of the candle was put off and small animals soon died i.e. it did not support either combustion or respiration. He named this gas in the retort 'azote' (nitrogen as it is known now) from the Greek word azotos meaning lifeless. He was able to demonstrate that solid calx (oxide of the metal) was formed due to the metal combining with an active part of air and not due to the loss of phlogiston. On the basis of this observation, he presented to the Academy on 26 April 1775, the famous memoir or paper (the Easter Memoir) titled *On the Nature of the Principle Which Combines with Metals during Their Calcination and Increases Their Weight* (Memoire

Sur La Combustion en General, Memoir on Combustion in General, *Mémoires de l'Académie Royale des Sciences* 1777, 592-600, From Leicester & Klickstein [8]).

Lavoisier continued his experiments on combustion by heating the calx of mercury as Priestley had done and again quantified the weight of the residue (of mercury) left in the container and the gas released and found its volume was 1/4 of its original volume. He noticed that a lighted "*taper burned in it with a dazzling splendour, and charcoal, instead of consuming quietly as it does in common air, burnt with a flame, attended with a decrepitating noise, like phosphorus, and threw out such a brilliant light that the eyes could hardly endure it. This species of air was discovered almost at the same time by Mr. Priestley, Mr. Scheele, and myself. Mr. Priestley gave it the name of dephlogisticated air*, *Mr. Scheele called it empyreal air. At first I named it highly respirable air, to which has since been substituted the term of vital air"* (From Lavoisier [5]).

He published his findings in the revised version of the Easter Memoir in 1778 where he stated that the part of air that combined with mercury to produce calx of the metal was "*nothing else than the healthiest and purest part of the air*" or the "*eminently respirable part of the air.*" (*'Considérations générales sur la nature des acides', "General Considerations on the Nature of Acids," 1778*) (From "Antoine Lavoisier" *Wikipedia* [9]). Based on the results observed from these two experiments, Lavoisier concluded that mercury combined with oxygen in air in the retort to form calx and when the calx was heated in another retort to a higher temperature, it decomposed into mercury and released the same volume of gas (oxygen) as had been consumed in the first experiment (this according to phlogiston theory was not possible). Priestley had observed the same phenomenon but explained it within the framework of the phlogiston theory.

$$\mathbf{Hg(l) + 1/2\ O_2(g) \longrightarrow HgO(s)}$$

$$\mathbf{HgO(s) \longrightarrow Hg(l) + 1/2\ O_2(g)}$$

Common air not an element: Lavoisier continued his study of combustion. Next he discovered that when five volumes of the residual air was added to one volume of the liberated air, the resultant mixture was similar to common atmospheric air in all respects. With this simple experiment, Lavoisier showed that common air was not an element as had been assumed for centuries, but a mixture of two distinct types of air with different chemical properties.

Naming of Oxygen: Lavoisier said *"I shall for the future call dephlogisticated air or eminently respirable air by the name of the acidifying principle, or, if the same meaning is preferred in a Greek word, by that of the oxygine principle"* (From Childs [1]).

Lavoisier observed that when non-metals like sulfur, phosphorus, charcoal, and nitrogen were burnt, the products were acidic. He wrongly concluded that the new liberated gas of his experiments was responsible for acidifying the products and consequently in 1789 he called this part of air, oxygen (oxy- meaning sharp and -gen meaning producing in Greek) attributing the sharp taste of acids to the presence of this part of air. Naming of dephlogisticated air (Priestley), or fire air or vital air (Scheele) as oxygen (in English) was a stroke of genius and is a perfect example of the synergy of Lavoisier's mastery of chemical phenomena, love of classics and poetry and command over language.

Henry Armstrong, an English chemist of the 19th century felt that "*In designing the word Oxygen, Lavoisier rose to the greatest height of his unparalleled genius. Not only is the word a monument to his astounding insight into chemical phenomena, to his philosophic power; it is also proof of a deep philological feeling and acumen, as well as of his sense of the beauty of words. Think of the astounding step he took, after his instant appreciation of Priestley's discovery, in translating the old nonconformist's ponderous reminder of the doubtful past of our science conveyed in the name Dephlogisticated Air into an all-significant word of the aural and lingual perfection of Oxygen .... think of him as the pioneer*

> *who not only sought to put system into the souls of chemists but also tipped their tongues with harmony*" (From Childs [1]).

Lavoisier was able to simultaneously a) disprove once and for all phlogiston theory that had survived for nearly a century and b) establish the true nature of combustion. He was also able to establish that air was not an element but a mixture of two gases oxygen and azote (nitrogen) and the total weight of substances before and after an experiment was constant (conservation of mass). These two experiments formed the basis for the Law of Conservation of Mass. On the basis of these experiments, Lavoisier gave comprehensive explanation of the phenomenon of combustion thus: *"In every combustion there is disengagement of the matter of fire or of light. A body can burn only in pure air [oxygen]. ..... the increase in weight of the body burnt is exactly equal to the weight of air destroyed or decomposed. The body burnt changes into an acid by addition of the substance that increases its weight. Pure air is a compound of the matter of fire or of light with a base. In combustion the burning body removes the base, which it attracts more strongly than does the matter of heat, which appears as flame, heat and light"* (From Lavoisier [6]).

Lavoisier made sure that he got the credit for enunciating the theory of oxidation and combustion. *"This theory is not, as I have heard it called, the theory of the French chemists in general, it is* ***mine****, and it is a possession to which I lay claim before my contemporaries and before posterity. Others, no doubt, have given it new degrees of perfection, but I hope that one will not be able to deny me the whole theory of oxidation and combustion; the analysis and decomposition of air by*

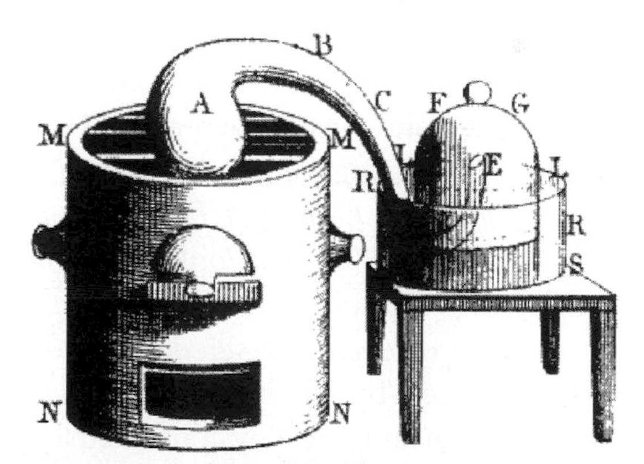

(From "Antoine Lavoisier", *Wikipedia*)

metals and combustible bodies; the theory of the formation of acids; more exact knowledge of a great number of acids, notably vegetable acids; the first ideas on the composition of plant and animal substances, and the theory of respiration" (From Childs [1]). In 1783, Lavoisier presented his famous paper *Réflexions sur le phlogistique* (Reflections on Phlogiston) to the French Academy in which he once for all demolished the phlogiston theory of combustion.

> "*If everything in chemistry is explained in a satisfactory manner without the help of phlogiston, it is by that reason alone infinitely probable that the principle does not exist; that it is a hypothetical body, a gratuitous supposition*" (From Lavoisier [6]).

Many senior chemists did not accept his revolutionary theory of combustion. Scheele expressed his rejection of Lavoisier's theory saying: *"Would it be so difficult to convince Lavoisier that his system of acids is not to everybody's taste? Nitric acid composed of pure air and nitrous air, aerial acid of carbon and pure air, sulphuric acid of sulphur and pure air!.... Is it credible? ... I will rather place my faith in what the English say" (Scheele to Bergman,* 1784) (From Childs [1]).

Eventually Lavoisier was proved right. He stated *"I see with great satisfaction that young people who begin to* (study) *the science without prejudice .....no longer believe in phlogiston in the way in which Stahl presented it, and look upon the whole of this doctrine as a scaffolding more embarrassing than useful for the continuance of the structure of chemical science"* (From Childs [1]). His theory was soon accepted by chemists who had earlier rejected it. Even though Cavendish and Priestley continued to believe in the phlogiston theory, Richard Kirwan of Ireland, Joseph Black of Scotland, Klaproth of Germany, Bergman of Sweden who had earlier rejected Lavoisier's idea and a majority of American and Russian soon accepted unequivocally the revolutionary explanation of combustion.

Joseph Black's letter to Lavoisier 1791

"The numerous experiments which you have made on a large scale, and which you have so well devised, have been pursued with so much care and with such scrupulous attention to details, that nothing can be more satisfactory than the proofs you have obtained. ......Having for thirty years believed and taught the doctrine of phlogiston as it was understood before the discovery of your system, I, for a long time, felt inimical to the new system, which represented as absurd that which I had hitherto regarded as sound doctrine; but this enmity, which springs only from the force of habit, has gradually diminished, subdued by the clearness of your proofs and the soundness of your plan" (From Childs [1]).

Respiration and combustion

"Respiration is a combustion, like that of candle burning" (From "Antoine Lavoisier" *Wikipedia* [9], Buchholz & Schoeller [10]).

"One would say that the analogy between respiration and combustion had not escaped the poets, or rather the philosophers of antiquity whose interpreters they were. The fire stolen from heaven, the fire of Promethium, is not merely an ingenious poetical idea; it is a faithful picture of the operations of Nature, at least for animals that breathe: we can therefore say with the ancients that the flame of life is lit at the instant the child draws its first breath and that it is extinguished only at death. ……. the fable is, indeed, only an allegory under which they (the ancient philosophers/scientists) hid the great truths of medicine and physics" (From Lavoisier [11]).

Lavoisier was a self-trained physiologist. Some regard Lavoisier's contribution to physiology does not get the attention it deserves. An important spin-off from Lavoisier's theory of combustion was the recognition that *"la respiration est donc une combustion (respiratory gas exchange is a combustion"* (From "Antoine Lavoisier" *Wikipedia* [9], Buchholz & Schoeller [10]). Even though the crucial role played by common air in both respiration and combustion was known, the commonality between the two processes was not clearly understood as one took place inside the animal body and the other in the open air. Lavoisier was interested in unraveling this common link and published a paper in 1777 titled *"Experiments on the respiration of animals; and on the changes effected on the Air in passing through their lungs"* on respiration based on the experimental data he had collected. However, he could arrive at the experimental proof of his belief that *"respiratory gas exchange is a combustion"* only by 1782/83. Lavoisier collaborated with Laplace and designed and built an ice calorimeter with two shells or layers. A live guinea pig was confined in the calorimeter and while the inner shell was filled with ice, the outer shell/layer was packed with snow to measure the amount of heat given off and the amount of carbon dioxide produced by the guinea pig during respiration in a fixed time frame. Next they burned sufficient quantity of carbon in the calorimeter to produce the same amount of carbon dioxide as the guinea pig exhaled and measured the quantity of heat generated. Lavoisier had succeeded in showing that *"la respiration est donc une combustion,"* that is, respiratory gas exchange is a combustion, like that of a candle burning. He also declared *"A man cannot live more than 24*

(From "Antoine Lavoisier" *Wikipedia*)

Madame Lavoisier's engraving of Lavoisier conducting an experiment on respiration.

hours unless he has at least three cubic meters of air that is being constantly replaced" (From Lavoisier [12]) Much later, he realized that heat generated by animals was not just due to respiration but it was also due to the changes that take place inside the cells with food consumed by the animal providing the fuel for this activity.

Water is a compound not an element

Lavoisier was aware (as many of his contemporaries were) that water may not be an element as it had been believed since antiquity. Lavoisier's discovery of the chemical nature of water follows a strikingly similar pattern to the discovery of oxygen. In 1781, as part of his experiments on combustion when he burnt inflammable air (discovered by Cavendish) in pure air or oxygen, he did not notice the formation of water even though inflammable air caught fire. As in the case of oxygen, the journey starts in England with Priestley. On heating a small quantity of calx of a metal in a tall flask containing "inflammable air", Priestley found a few drops of dew on its walls. On hearing about this experimental result, Cavendish repeated the experiment and arrived at the same result, but realised that the dew was actually water. As both of them believed in the phlogiston theory, they concluded that water or dew was formed when inflammable gas and dephlogisticated air lost phlogiston that was present in them and they lost interest in further investigation. Blagden, once a secretary of Cavendish, who was visiting Paris gave the details of Cavendish's experiments to Lavoisier and other chemists. Lavoisier at once grasped the true implications of the experiment. Here, it was an experiment on synthesis while the earlier one had been on decomposition – in both cases oxygen and heat played crucial roles. Lavoisier and Laplace almost with indecent haste conducted the experiment by burning the two gases in a closed container and synthesized water in Blagden's presence on June 24, 1783 and reported their findings in a memoir to the French academy the next day. He christened inflammable gas of Cavendish, hydrogen (meaning water-former in Greek). Next, Lavoisier decomposed water into oxygen and inflammable air to make doubly sure that water was indeed a compound of oxygen and hydrogen (the inflammable gas). He also found that water was not an element, not a simple substance as believed. It was composed of 14,338 parts of oxygen and 85,688 parts of hydrogen (From Aykroyd [13]). At last Lavoisier's view on combustion and the role of oxygen in the process was accepted and phlogiston evaporated into thin air permanently.

> As in the case of oxygen, Lavoisier either by oversight or deliberately did not acknowledge the contributions of Priestley and Cavendish. This omission proved costly as many scientists were scathing in their criticism and some went to the extent of accusing him of plagiarism. Marat who was responsible for preparing the charge sheet against Lavoisier later accused him of plagiarism.

Many agreed with Justus von Liebig, who wrote in Letters on Chemistry, No. 3, *"Lavoisier discovered no new body, no new property, no natural phenomenon previously unknown; but all 'the facts established by him were the necessary consequences of the labours of those who preceded him. His merit, his immortal glory, consists in this—that he infused into the body of the science a new spirit; but all the members of that body were already in existence, and rightly joined together"* (From Liebig [14]).

Classification and nomenclature of elements

"Every branch of physical science must consist of three things: the series of facts which are the objects of the science, ideas which represent these facts and the words by which these idea are expressed. Like three impressions of the same seal, the word ought to produce the idea and the idea to be a picture of the fact".

"As ideas are preserved and communicated by means of words, it necessarily follows that we cannot improve the language of any science, without at the same time improving the science itself; neither can we, on the other hand, improve a science without improving the language or nomenclature which belongs to it" (From Lavoisier [5]).

Lavoisier's contribution to systematizing chemical nomenclature is as profound as his contribution to the understanding of combustion and respiration and had equally far reaching consequences in revolutionizing the subject. Before Lavoisier, the language of chemistry was imprecise, ambiguous and the terms that were used had Greek, Latin and Arabic origins. Language of chemistry was confusing as the same substance had many different and unsystematic names with relation to the chemical composition, for example Epsom Salt, Fuming Liquor of Libavius, Butter of Arsenic, Oil of Vitriol (for sulfuric acid), 'flower of zinc' (for zinc oxide), 'Spanish Green' (for copper acetate). Lavoisier felt strongly that chemistry needed a language that could express clearly the chemical concepts and to do this the names of chemical substances should indicate not only the elements in them but also their proportion present in that particular substance. Therefore, he along with Berthollet (and Fourcroy to a lesser extent) devised a system of chemical nomenclature based on the following basic principles! (a) Name of a substance must have universal acceptance, (b) Composition of the substance must be reflected by the name wherever possible, (c) Foreign names must be of Greek or Latin origin. Lavoisier strictly followed these tenets and established the method for naming of chemical elements and substances (for example sulfurous acid contains less oxygen than sulfuric acid). In 1787, Lavoisier and his collaborators published *"Methode de Nomenclature Chimique"* (Methods of Chemical Nomenclature). With its publication, Lavoisier et al replaced the haphazard method of naming chemical substances that had been in use for centuries with a well constructed system of chemical nomenclature which has remained relatively unchanged till today. As in the case of his ideas on combustion, many contemporary chemists did not accept this method of nomenclature. But the consensus among chemists now is that even if Lavoisier had done nothing else, this novel system of chemical nomenclature would have been enough for him to be regarded as one of the chemical pioneers.

Lavoisier went on to classify the then known elements under the following four groups:

1. Elastic fluids: Under this category Lavoisier included oxygen, nitrogen, hydrogen as well as light, heat.

2. Nonmetals - Group of "oxidizable and acidifiable non-metallic elements". Lavoisier classified sulfur, phosphorus, carbon, hydrochloric acid, hydrofluoric acid, and boric acid as oxidizable and acidifiable nonmetals.
3. Metals: Under this group Lavoisier listed the following as the elements that are "metallic, oxidizable, and capable of neutralizing an acid to form a salt." Antimony and arsenic, silver, bismuth, cobalt, copper, tin, iron, manganese, mercury, molybdenum, nickel, gold, platinum, lead, tungsten, and zinc. Today antimony and arsenic are not considered as metals.
4. Earths: Lavoisier classified lime, magnesia, baryta or barium oxides, alumina and silica as earthy elements in the solid state which formed salts.

Author and a polymath

Lavoisier consolidated all his ideas in his book *Traite Elementaire de Chimie published* in 1790. Many chemists consider this as the first authentic text book. The enormous influence this book had on contemporary chemistry of his time and the future of chemistry made many to compare the far reaching influence this book had on the evolution of chemistry to that of Newton's *Principia* on physics.

Lavoisier the social activist: Lavoisier was a committed social reformer. He brought in agricultural reforms, advocated scientific farming to increase production, based on his discovery of the process of respiration in animals; he fought for reforms in hospitals, recommended positive actions to improve the deplorable conditions of the prisons.

It is truly amazing that Lavoisier achieved all this by working for just a few hours a day at his home laboratory before and after his official work. It is said that he worked at chemistry from 7 am to 9 am and again in the evening after dinner from 9 pm to 11 pm and devoted entire Saturday to work with his students.

'We all teach ... the chemistry of Lavoisier and Gay-Lussac'. Marcellin Berthelot (1877) (From Crosland [15]).

Lavoisier left a rich and mindboggling legacy for future generations. By emphasizing the importance of experimental verification of every hypothesis based on accurate measurement, he changed the way chemistry functioned.

Lavoisier predicted that *"Perhaps... some day the precision of the data will be brought so far that the mathematician will be able to calculate at his desk the outcome of any chemical combination, in the same way, so to speak, as he calculates the motions of celestial bodies"* (From Heilbron [16]).

Science and society: Lavoisier understood the crucial role played by a scientist in nation building. He believed that *"It is not absolutely necessary, in order to deserve well of mankind and to do one's duty to one's country, to be called to glittering public office for the organization and regeneration of empires. The man of science in the silence of his laboratory and his study can also serve his country: by his work he is enabled to hope that he may diminish the sum of the evils that afflict the human race, and increase enjoyment and happiness; and were he to contribute, by the charting of new ways, only to the prolongation by some years, even by some days, of the average life of men, he could aspire also to the glorious title of benefactor of humanity"* (Lavoisier, Memoir 1789).

The final days

In spite of his life-long commitment to social justice and improving the living conditions of the poor and his unrivalled contributions to chemistry, it was his closeness to the royalty and his work at the Ferme Generale that proved fatal. Lavoisier saw the opposition to the rich and powerful steadily growing towards the end of the 1780s, and by 1789 the rumblings had grown into a revolt by the populace of France, and especially Paris, against the rich and the powerful. *L'Academie Royale des Sciences* founded in 1666 by Grand monarch Louis XIV and all other Academies became defunct. During the turbulent times of the French Revolution a reign of terror swept across France and the tax collectors were hated for their alleged corrupt practices. *After making all efforts to avoid being arrested by the Revolutionary Tribunal for his membership of the Farmers-general, Lavoisier was finally arrested, together with his father-in-law, and put in prison on the 28th November 1793. He and his fellow farmers-general had been arrested on allegations of financial misconduct without clinching evidence and on the spurious charge of selling tobacco with too much water* (From Childs [1]). Even though Lavoisier was scrupulously honest in his dealings, he was in the company of some dishonest and unscrupulous people and he was tried as a traitor. Lavoisier had made many enemies with his outspoken and uncompromising attitude to intellectual pretenders. Also his wealth, fame, efficiency in doing his job whether it was for the people or in pursuing chemistry and his lavish living at his quarters at the Arsenal added to his woes. One of Lavoisier's enemies was Jean Paul Marat who hated Lavoisier ever since Lavoisier had dismissed his claims of a scientific discovery. Marat had become a influential revolutionary. One of Lavoisier's so-called crimes was that he had obtained for some of his foreign-born scientist friends, exemption from the harsh mandate of stripping all foreigners born in enemy countries of freedom and right for personal possessions. Lagrange, the German mathematician, was one of those who benefitted from Lavoisier's intervention. In the end, his unselfish work for improving the lot of the poor counted for nothing. Soon after his arrest, the tribunal confiscated Lavoisier's property. Lavoisier remained dignified and defiant. He wrote to his cousin Augez de Villiers just before his end,

"I have had a fairly long life, above all a very happy one, and I think that I shall be remembered with some regrets and perhaps leave some reputation behind me. What more could I ask? The events in which I am involved will probably save me from the troubles of old age. I shall die in full possession of my faculties, and that is another advantage that I should count among those that I have enjoyed. If I have any distressing thoughts, it is of not having done more for my family; to be unable to give either to them or to you any token of my affection and my gratitude is to be poor indeed" (Lavoisier in a Letter to Augez de Villiers, From McKie [4]).

Lavoisier was tried, convicted and was taken through the streets of Paris in a dilapidated open cart to the place designated for being guillotined on May 8, 1794 in Paris. His severed head and the headless body was carted away and thrown along with all the executed victims and unceremoniously dumped into a ditch dug up for this purpose in Errancis cemetery. At Lavoisier's untimely and unnecessary death, Lagrange paid his immortal tribute *"It took them only an instant to cut off his head, but France may not produce another such head in a century"* (From Guerlac [17]). After his execution, Lavoisier's personal possessions including his notebooks, papers and books were in the process of being destroyed. Fortunately, the period of madness passed and within 18 months of his death, Marie Lavoisier was officially informed in 1795 about absolving Lavoisier of all the charges, and his property and belongings were returned to her.

"Thus died France's greatest scientist whose labours brought chemistry a new and logical theory, a new and systematic nomenclature, a fresh and scientific outlook, and who had swept away all vestiges of alchemy and superstition. In addition to this remarkable achievement his pioneering work in agriculture, economics, finance, politics, sociology, education, and hygiene all bore the indelible stamp of his impeccable logic, keen practical sense, and strong appreciation of the urgent needs for constructive changes in these fields" (From Duveen [18]).

Three years after his untimely death, in August 1796, *Lycee de Arts* organized a formal funeral to Lavoisier though without his mortal remains. A magnificent banner with words "A l'immortel Lavoisier" hung across the portal of the Lycee. The darkened hall inside was decorated with wreaths and twenty lamps and a massive chandelier. Amidst tributes from the eminent people of the new Republic, Lavoisier was finally laid to rest.

References

1. Peter E. Childs, *The life and legacy of Antoine-Laurent Lavoisier*, 1994. (http://www.ul.ie/~childsp/CinA/Issue43/editorial43.html) (From the Website Antoine-Laurent Lavoisier, Chemistry Materials from Ray Tedder).

2. E. J. Holmyard, *Chemistry to the time of Dalton,* Oxford University Press, Oxford 1925 (From the Website Antoine-Laurent Lavoisier, Chemistry Materials from Ray Tedder).

3. Website: Lavoisier, Antoine (1743-1794) (scienceworld.wolfram.com) (http://scienceworld.wolfram.com/biography/Lavoisier.html).

4. Douglas McKie, *Antoine Lavoisier: Scientist, Economist, Social Reformer* (1952).

5. Antoine-Laurent Lavoisier, *Elements of Chemistry* (1790), trans. R. Kerr.

6. Lavoisier, "Réflexions sur le phlogistique', *Mémoires de l'Académie des Sciences*, 1783, 505-38. Also Reprinted in *Oeuvres de Lavoisier* (1864), Vol. 2, 640, trans. M. P. Crosland.

7. *Oeuvres de Lavoisier, Correspondance*, Fasc. II. 1770-75 (1957), 389-90. From *The Eighteenth-Century Revolution in Science,* translation by A. N. Meldrum, (1930).

8. Henry Marshall Leicester and Herbert S. Klickstein, *A Source Book in Chemistry 1400-1900* (1952). (Memoire Sur La Combustion en General, Memoir on Combustion in General, *Mémoires de l'Académie Royale des Sciences* 1777, 592-600).

9. "*Considérations générales sur la nature des acides*" ("General Considerations on the Nature of Acids,") 1778, "Antoine Lavoisier." *Wikipedia, The Free Encyclopedia*. Wikipedia, 25 Feb. 2015.

10. Andrea C Buchholz and Dale A Schoeller, *Is a calorie a calorie? Am J Clin Nutr 2004 79: 899S-906S* ("Antoine Lavoisier." *Wikipedia, The Free Encyclopedia*. Wikipedia, 26 Feb. 2015).

11. Lavoisier, Memoir (1789), *(from the website Antoine-Laurent Lavoisier, Chemistry Materials from Ray Tedder).*

12. Website: Antoine-Laurent de Lavoisier (1743-1794), 3) Lavoisier, Pioneer in Public Health, (*historyofscience.free.fr/**Lavoisier**-Friends/ a_chap3_**lavoisier**.html)* Lavoisier, *Oeuvres*, vol. III, p.647.

13. W. R. Aykroyd, *Three Philosophers: Lavoisier, Priestley and Cavendish* 2014, p84.

14. Justus von Liebig, (ed. David M. Knight), *The Development of Chemistry, 1789-1914: Familiar letters on chemistry,* Letter III, Routledge, 1851, reprinted 1998.

15. Maurice P. Crosland, *Gay-Lussac: Scientist and Bourgeois*, Cambridge University Press, 1978.

16. J. L. Heilbron,Weighing imponderables and other quantitative science around 1800, University of California Press, 1993.

17. Henry Guerlac, *Antoine-Laurent Lavoisier – Chemist and Revolutionary*. Scribner (1975).

18. Denis I. Duveen in *Great Chemists* Interscience, New York and London 1961. (from the Website Antoine-Laurent Lavoisier, Chemistry Materials from Ray Tedder).

CNR Rao, *Understanding Chemistry*, Universities Press (1999).

Portrait of Monsieur de Lavoisier and his Wife, chemist Marie-Anne Pierrette Paulze by Jacques-Louis David, 1788, Source: Metropolitan Museum of Art. (From Antoine Lavoisier, From Wikipedia, the free encyclopedia).

Antoine Lavoisier's famous phlogiston experiment. Engraving by Mme Lavoisier in the 1780s taken from Traité élémentaire de chimie (Elementary treatise on chemistry) Source: Hidrogenexp2.gif, Modifications made by Cdang. (From Antoine Lavoisier, From Wikipedia, the free encyclopedia).

Lavoisier conducting an experiment on respiration in the 1770s, drawing by Madame de Lavoisier. "Lavoisier humanexp". Licensed under Public Domain via Wikimedia Commons - http://commons.wikimedia.org/wiki/ File:Lavoisier_humanexp.jpg#mediaviewer/File:Lavoisier_humanexp.jpg (From Antoine Lavoisier, From Wikipedia, the free encyclopedia).

2. JOHN DALTON (1766–1844)

Proponent of the concept of the atom

If, in some cataclysm, all of scientific knowledge were to be destroyed and only one sentence passed on to the next generation of creatures, what statement would contain the most information in the fewest words? I believe it is the atomic hypothesis... that all things are made of atoms....In that one sentence... there is an enormous amount of information about the world, if just a little imagination and thinking are applied" - Richard Feynman (From Feynman et al. [1])

(From "John Dalton" *Wikipedia*)

It was Dalton who first showed the scientific world that atoms, far from being just an abstract philosophical concept prevalent in ancient India and Greece, was indeed the basic constituent of all matter. In his "A New

System of Chemical Philosophy", he says, *"Matter, though divisible in an extreme degree, is nevertheless not infinitely divisible. That is, there must be some point beyond which we cannot go in the division of matter. The existence of these ultimate particles of matter can scarcely be doubted, though they are probably much too small ever to be exhibited by microscopic improvements. I have chosen the word atom to signify these ultimate particles..."* (From Freund [2]).

Early life

John Dalton was born in 1766 in a poor Quaker family in England. His father Joseph Dalton was a weaver. Even though his mother Deborah Greenup Dalton came from a slightly more affluent background, the family could not afford to provide John and his brother formal education. They received only basic training in the three Rs in Pardshaw Hall, a Quaker grammar school in Eaglesfield. (Even this was a luxury as only one in 200 in that area was literate!). John became a teacher at the age of 12 at this school, after his elder brother became the owner of the school. Two years later, the brothers bought a school at Kendal. While studying Young John came under the influence of Elihu Robinson (a well to do Quaker in Eaglesfield) and John Gough, visually impaired son of a rich merchant in Kendal. He was encouraged to develop interest in meteorology, mathematics and science. Perhaps the most important knowledge he gained from them was how to make and use simple meteorological instruments and to keep an accurate and detailed log of daily weather conditions. Dalton lived in the ideal place to pursue this interest as the weather in the Lake Districts changed constantly. Measuring the temperature and humidity at different altitudes meant actually climbing mountains in the surrounding areas. Dalton started keeping a daily log of accurate meteorological data in 1787 (when he was 21 years old) and maintained it almost till the last day of his life. It is said that his log contained observations in excess of 200,000 entries (From Smith [3]). Perhaps this interest in the daily changes in the behaviour of gases in air led to Dalton's lifelong interest in this subject. Incidentally this meticulous observation of the behaviour of gases in atmospheric air contributed to his generalizations about atoms.

Based on his extensive observations, contrary to the belief then that the atmosphere was a compound of specific elements, Dalton surmised correctly that the it was a mixture consisting approximately 80 percent nitrogen and 20 percent oxygen. He also measured atmosphere's capacity to absorb water vapour and recorded how its partial pressure varied with temperature. Dalton is often called the father of meteorology for changing meteorology from a subject of popular folklore to a serious study of climatology.

Dalton joined New College in Manchester in 1793 as a teacher of mathematics and natural philosophy. One of the first things Dalton did was to join the Manchester Literary and Philosophical Society. Within a short time after joining the Society, on 31 October 1794, Dalton presented his first paper on colour blindness titled *"Extraordinary facts relating to the vision of colours"*. He hypothesised in the paper that the defect of not seeing certain colours was due to the discoloration of the liquid medium of the eyeball. Until Dalton's presentation, there was little awareness that colour blindness was a serious problem. Dalton also realised that it could be a hereditary problem as he and his elder brother both suffered from it. He described his condition in the essay "*That part of the image which others call red, appears to me little more than a shade or defect of light; after that, the orange, yellow and green seem one colour, which descends pretty uniformly from an intense to a rare yellow, making what I should call different shades of yellow*" (From Dalton [4]).

Dalton bequeathed his eyes for further studies on the problem. On investigation of the remains of the eye, it was found that Dalton's theory was wrong.

Dalton published his first book titled "*Meteorological Observations and Essays*" within a short period of joining the Society. He conducted a number of experiments which enabled him to arrive at an understanding of the behaviour of gases in mixtures. He published his conclusions (four essays) in "*Experimental Essays on the Constitution of Mixed Gases; on the Force of Steam or Vapour from water and other liquids in different temperatures, both in a Torricellian vacuum and in air; on*

Evaporation; and on the Expansion of Gases by Heat". He observed that when two or more gases were mixed together, they did not repel or attract each other. The result of this independent behaviour was that each gas exerted its own pressure. The result of this "independence" was that the total pressure exerted by the mixture of gases was the sum of the separate pressures exerted independently by the constituent gases. He was able to show that the pressure exerted by a vapour and its ambient temperature were mathematically linked. His studies also led him to observe that when certain gases formed a mixture, their ratio remained constant irrespective of the amount or quantity of the mixture thus formed. He rightly surmised that this was because the ratio was maintained to the smallest particle of each constituent gas. Dalton conducted many experiments to verify his observation and arrived at the conclusion that he could calculate the weight of the smallest particle of each element. Based on his experimental findings, Dalton enunciated the Law of Partial Pressures.

Law of Multiple Proportions and the beginning of Atomic Theory

In 1803, Dalton began conducting experiments in which he reacted nitric oxide with oxygen. He was puzzled to find that when the proportions of the two gases or the ratios of the combining gases (nitric oxide and oxygen) were changed, the composition of the resultant gas formed was different. Under one set of conditions nitrogen of nitric oxide combined with oxygen in the ratio 1:1.17 and under different set of conditions, it combined with oxygen in the ratio 1:1.3. Further experiments led him to arrive at the generalization that the elements combined with one another (to form compounds) where the ratio of their weights was always whole numbers - the basic premise of The Law of Multiple Proportions.

$$\mathbf{2NO + O \longrightarrow N_2O_3}$$

$$\mathbf{NO + O \longrightarrow NO_2}$$

This observation paved the way for Dalton's work on developing a table of atomic weights. To compile the table, he took the weight of hydrogen as 1 (as it was the lightest element known then) and as the standard weight against which the weights of other elements were compared.

Birth of the idea of an atom: The idea of an atom seems to have occurred to Dalton by the physical behaviour of gases. In his paper on absorption of gases published in 1805, he says, *"why does not water admit its bulk of every kind of gas alike? This question I have duly considered, though I am not able to satisfy myself completely, I am nearly persuaded that the circumstance depends on the weight and number of the ultimate particles of the several gases"* (From Henry [5]).

The next problem that Dalton faced was how to express these purely mental images of his revolutionary discovery verbally and pictorially so as to make them easy to comprehend. Dalton found the answer to the vexing problem in an unlikely source - his extensive study of ancient philosophers and their views on matter. They included ideas of Kanaada of India who used the term "paramanu" to describe the smallest particle of matter and of Democritus and Leucippus of Greece who used the Greek word atomos (meaning that which cannot be cut) to describe the smallest indivisible particle of matter and said that all matter consisted of atoms which moved through the empty space separating them. Newton's view (. . . *It seems probable to me, that God in the beginning formed matter in solid, massy, hard, impenetrable, moveable particles, of such sizes and figures, and with such other properties, and in such proportion to space, as most conduced to the end for which he formed them; and that these primitive particles……….are incomparably harder than any porous bodies compounded of them; ……no ordinary power being able to divide what God himself made one in the first creation"*, From Newton [6]), gave Dalton the solution to the verbal part of problem – how to define an abstract concept. Putting together the essence of these ideas, Dalton came to the conclusion that atom of one element differed from atom of any other element only by its weight. As a result of this insightful understanding, Dalton's Atomic Theory was born.

Dalton's Atomic Symbols: *"Another immense service rendered by Dalton, as a corollary of the new atomic doctrine, was the creation of a system of symbolic notation, which not only made the nature of chemical compounds and processes easily intelligible and easy of recollection, but, by its very form, suggested new lines of inquiry"* (From Huxley [7]).

Dalton still had to solve the problem of how to represent atoms pictorially so that these invisible and minute particles could be "seen" and manipulated. He did not agree with Berzelius' symbols for atoms. He felt that *"A young student in chemistry might as soon learn Hebrew as make himself acquainted with them........ They appear to me equally to perplex the adepts in science, to discourage the learner, as well as to cloud the beauty and simplicity of the atomic theory"* (From Ihde [8]).

Dalton found the solution by using circles with distinguishing features inside the circle to represent the atom of a particular element. A sample representation is given below.

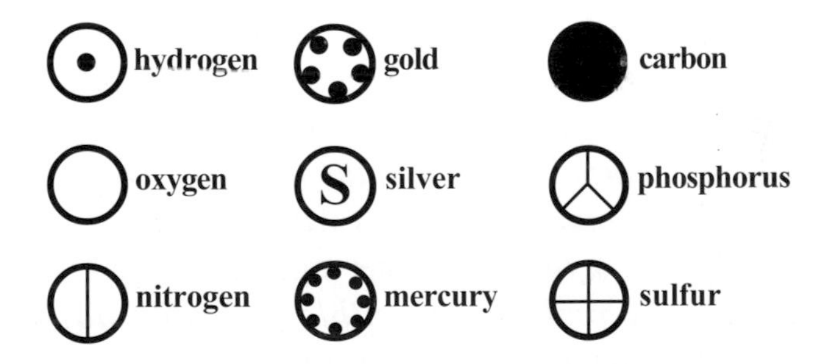

By this method of representing invisible atoms, Dalton had created a laboratory on paper where the atoms could be manipulated and compounds could be created and later they could be verified in an actual laboratory - early days of modern day computerized model building!

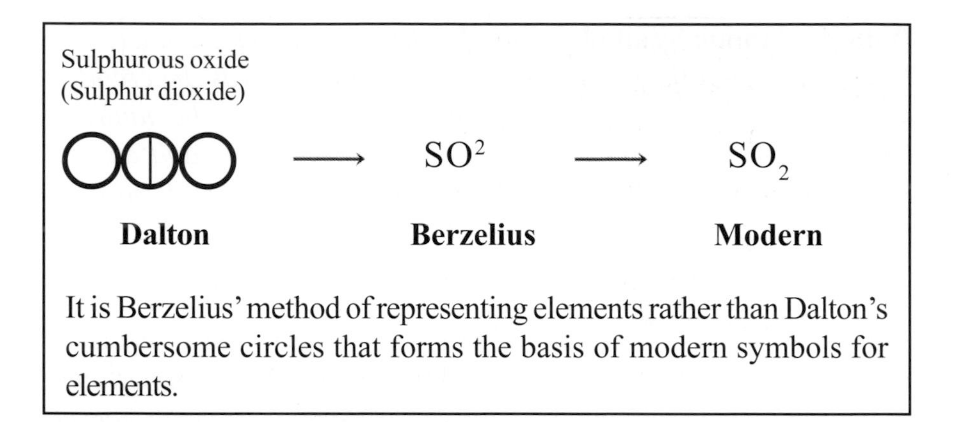

It is Berzelius' method of representing elements rather than Dalton's cumbersome circles that forms the basis of modern symbols for elements.

Dalton announced his table of relative weights of atoms on October 21, 1803 at the Manchester Literary and Philosophical Society. The scientific world woke up to the world of atoms. This fundamental discovery immediately caught the attention of The Royal Institution of London. He was invited to present his revolutionary theory to a distinguished audience, but not everyone there was enthusiastic about it. Influential chemists including Humphry Davy were hostile to the idea. Davy went to the extent of dismissing Dalton's theory as *"rather more ingenious than important"* (From nitum.wordpress.com [10]) and dismissed Dalton's observation that *"chemical analysis and synthesis go no farther than to the separation of particles one from another, and their reunion. No new destruction or creation of matter is within the reach of chemical agency"* (From Dalton [9]), with the counter argument *"There is no reason to suppose that any real indestructible principle has yet been discovered"* (From nitum.wordpress.com [10]).

Dalton was often mocked for his lack of sophistication. He returned to Manchester a sad and disappointed man. However, he had the last laugh when after more experimental work, decades later, even his detractors including Humphry Davy had to admit that Dalton was right in his conjecture - that the basic constituent of all matter was atoms. Humphry Davy finally acknowledged Dalton's contribution with this compliment by comparing Dalton's seminal atomic theory to Kepler's contribution to astronomy.

"*The ultimate particles of all homogeneous bodies are perfectly alike in weight, figure &c*" (From Dalton [9]).

The main axioms of Dalton's Atomic Theory are:

1. All elements (or matter) are made up of atoms, the smallest particle.
2. All atoms of a particular element are identical in all respects including their chemical properties and atoms of one element cannot be changed into atoms of a different element.
3. Atoms of different elements can be identified by their weight.
4. In a chemical reaction, atoms of different elements always combine in fixed ratios to form new chemical compounds. In the process, they combine, or are rearranged.
5. Atoms cannot be further divided and can be neither destroyed nor created. It must be noted that not all of them have survived to the modern day.

Dalton compiled a list of 20 elements and using his symbols for them pictorially represented the atomic structure of seventeen simple molecules.

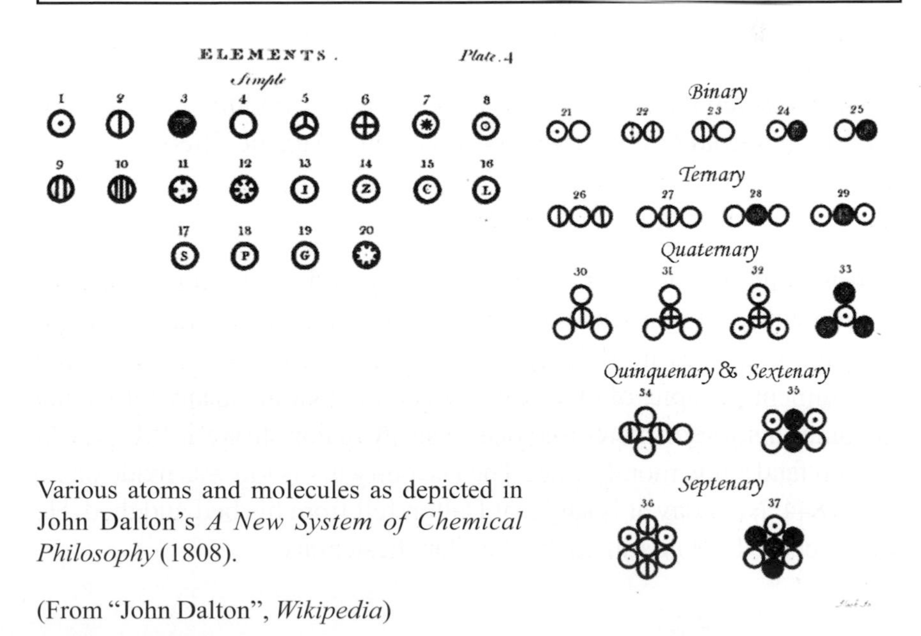

Various atoms and molecules as depicted in John Dalton's *A New System of Chemical Philosophy* (1808).

(From "John Dalton", *Wikipedia*)

In his book A *New System of Chemical Philosophy, (1808), Vol. 1, 212-3.* Dalton explained the manner in which atoms of different elements combined thus

"If there are two bodies, A and B, which are disposed to combine, the following is the order in which the combinations may take place, beginning with thc most simple: namely,

1 atom of A + 1 atom of B = 1 atom of C, binary
1 atom of A + 2 atoms of B = 1 atom of D, ternary
2 atoms of A + 1 atom of B = 1 atom of E, ternary
1 atom of A + 3 atoms of B = 1 atom of F, quaternary
3 atoms of A and 1 atom of B = 1 atom of G, quaternary" (From Dalton [9]).

Dalton the man

Dalton remained single and had only a few friends. He lived a simplc life of a typical Quaker. He was a physicist, meteorologist and a teacher and contributed essays on diverse topics in these subjects to the Literary and Philosophical Society of Manchester but his greatest legacy is his Atomic Theory.

He was elected a fellow of the Royal Society, London, and received the Royal Medal of the Royal Society for his Atomic Theory.

Last days

Dalton continued to work in his laboratory and teach mathematics to private students to earn his keep. His health started to fail and he suffered two minor strokes in 1837 and 1838 (which left him with a speech impediment). In spite of this, he continued to work in his laboratory and maintain his log of meteorological data. A major stroke in May 1844 proved fatal a few months later. The last entry in his log was made on 26 July 1844. Next day, it is said that Dalton fell from his bed and died. He was accorded a state funeral by the Manchester city.

References

1. Richard P. Feynman, Robert B. Leighton, Matthew Sands, *The Feynman Lectures on Physics, Volume 1, The Definitive Edition*, Sec; 1-2 Matter is made of atoms, chapter 1, Atoms in motion, Dorling Kindersley India Pvt. Ltd., 2009.
2. Ida Freund, *The Study of Chemical Composition: An Account of its Method and Historical Development*, Cambridge University Press, 1904, p 288.
3. Robert Angus Smith, *Memoir of John Dalton and History of the Atomic Theory*. (1856), London: H. Bailliere. p. 279.
4. John Dalton,*"Extraordinary facts relating to the vision of colours" Paper presented to Literary and Philosophical society, Manchester* 31 October 1794.
5. William Charles Henry, *Memoirs of the Life and Scientific Researches of John Dalton*, London 1854. P58 (Quoted from the concluding paragraph of the paper 'The absorption of gases by water and other liquids' (1803).
6. Sir Isaac Newton, *Opticks,* Query 31, Fourth Edition, 1730. (From Isaac Newton, Wikiquote, the free Encyclopedia , 24 February 2015).
7. T. H. Huxley, *The Advance of Science in the last half-century*, 1889.
8. *Aaron J. Ihde, The Development of Modern Chemistry* p 115, 1970.
9. John Dalton, *A New System of Chemical Philosophy* (1808), Vol. 1.
10. Biography of Davy, Sir Humphry, Baronet, September 28, 2012, https://nitum.wordpress.com

British physicist and chemist John Dalton (1766-1844) by Charles Turner (1773-1857) after James Lonsdale (1777-1839), 1834 (From John Dalton, *Wikipedia, The Free Encyclopedia*. 13 Apr. 2015.

"A New System of Chemical Philosophy fp" by haade - En.wiki. Licensed under Public Domain via Wikimedia Commons - http://commons.wikimedia.org/wiki/File:A_New_System_of_Chemical_Philosophy_fp.jpg#/media/File:A_New_System_of_Chemical_Philosophy_fp.jpg.

3. HUMPHRY DAVY (1778–1829)

The great discoverer and showman

"It may be said of modern chemistry, that its beginning is pleasure, its progress knowledge, and its objects truth and utility" (From Davy [1]).

*"Humphry Davy used his chemical discoveries, his wildly popular lecture series, and his general writings on science, to turn the "Chemical Philosopher" (*the term *scientist* not being coined until 1834*) into a figure of social and cultural importance in a quite new way"* (From Holmes [2]).

(From commons.wikimedia.org)

"Fortunately science, like that nature to which it belongs, is neither limited by time nor by space. It belongs to the world, and is of no country and of no age. The more we know, the more we feel our ignorance; the more we feel how much remains unknown....there are always new worlds to conquer." - Discourse Delivered by Davy at the Royal Society November 30, 1825 (From Davy[3]).

"It (science) has bestowed on him (a scientist) powers which may be almost called creative; which have enabled him to modify and change the beings surrounding him, and by his experiments to interrogate nature with power, not simply as a scholar, passive and seeking only to understand her operations, but rather as a master, active with his own instruments" (From Holmes [4]).

Early years

Humphry Davy, a contemporary of Lavoisier of France, shared Lavoisier's view of science as well as his passion for experiments. Like Lavoisier, he was a many-faceted genius. In contrast to Lavoisier's privileged background, Davy was the son of a wood carver in Cornwall, England and did not have the benefit of studying in a privileged school which exposed him to liberal education and to scientific subjects.

Humphry Davy, born on December 17, 1778, was the eldest child of Robert Davy who chose wood carving as a profession and Grace. Robert Davy lost his modest earning by bad investment in the coal mines and the family had to face dire financial problems after his death. Humphry Davy was a precocious child, encouraged by his doting family who took delight in his prowess. Even at a tender age while apparently just turning the pages of a book, he was actually reading it rapidly and could recall accurately what he read. As Davy remarked later, "*I had scarcely attained my fifth year before letters and words were perfectly at my command........I had learnt to associate enjoyment with praise"* (From Fullmer [5]). Like most children he was fond of listening to stories but unlike most, he could repeat them verbatim after listening to it once. According to his friends, he retained this gift in his later years also. At a time when men who indulged in scientific pursuits came from wealthy families and had the advantage of holistic education, Humphry Davy was a self educated man who set goals for himself and worked towards achieving them. He started his education in a local school at Penzance and later at a grammar school in Truro, Cornwall, where he did not display any interest in scientific pursuits or any drive or interest to acquire knowledge. Of his education, Davy himself said: *"I consider it fortunate, I was left much to myself as a child, and put upon no particular plan of study and that I enjoyed much idleness at Coryton's school. What I am, I made myself. I say this without vanity and in pure simplicity of heart"* (From Fullmer [5], Paris [6]). He was interested in reading historical stories and fantasies like the Arabian Nights and Aesop's Fables and John Bunyan's Pilgrim's Progress was one of his favourite books. In fact, his main interest at that time was in storytelling and in translating verses from classics. Even as a boy of eight he was a great story teller. Young boys would gather in the market square, Humphry would climb any cart parked there and from his exalted position regale his audience with stories from his latest readings. He came to be known as a *tale teller*. Humphry found great happiness from this and recalled later that *"applause of my companions, was my recompense for punishments incurred for being idle"* (From Paris [7]). When he turned 14, he joined the grammar school at Truro where Reverend Cardew was the headmaster. His stay in Truro Grammar School was so unremarkable that the principal of the school wrote to Davies Gilbert (one of Davy's staunch supporters) *"I could not observe any propensity to those scientific pursuits which raised him to such eminence. His best exercises were translations from the classics into English verse"* (From Fullmer [5]). Davy did not enjoy the regimented approach at the school and was relieved to leave the school at Truro. He wrote to his mother *"I was rejoiced when I first went to Truro school; but I was much more rejoiced when I left it forever. Learning naturally is a true pleasure: how unfortunate then it is that in most schools it is made a pain?"* (From Davy [8]).

When his family shifted to Verlace, young Humphry stayed back in Penzance with John Tonkin, surgeon-apothecary. He enjoyed the freedom there as he often pinched chemicals from the shop to make fire crackers and other compounds that gave out light. He improvised using Dr. Tonkins' tobacco pipes, tea cups from the kitchen to fabricate his apparatus and his eldest sister Kitty who doubled up as his assistant fetching things for Davy to conduct his experiments, bitterly complained about her dresses ruined by corrosive substances used by Davy in his experiments. Davy's interest in doing science experiments was further kindled by one Mr. Robert Dunkin who became Davy's friend and mentor. Mr. Dunkin was a saddler by profession but was interested in building scientific models to illustrate scientific principles. He used these to teach Davy the rudiments of experimental science. Davy started his future career as a great experimental chemist in Dunkin's house. Of his experiments, his friends often said: *"This boy Humphry is incorrigible. He will blow us all"* (From Hunt [9]).

Davy would often argue with Dunkin on a variety of topics. One winter day, to prove his point that heat was not a material, he took Dunkin to the frozen Larigan river in Penzance and demonstrated that rubbing motion between two pieces of ice produced enough energy to melt them and once the motion was stopped they fused together. Though he did not understand the principle at that time, he repeated the same experiment at the Royal Institute to illustrate the concept of cold heat which impressed the gathering there. As professor at the Royal Institution, Davy would later repeat many of the ingenious experiments which he learned from his friend and mentor, Robert Dunkin.

Davy enjoyed roaming along the rugged coast of Penzance writing poetry, fishing and collecting minerals. His carefree days ended rather abruptly when his father died in 1794 leaving his mother heavily in debt, with no resources to take care of herself and her five children. Humphry Davy as the eldest son had to give up school to assume responsibilities of a sole bread winner. Mr. Tonkin, family friend and godfather, took Davy under his wing and apprenticed him to surgeon J. Bingham Borlase at Penzance. During his apprenticeship, young Humphry decided to take up medicine as a career and to prepare himself for it, he embarked on rigorous and systematic self-instruction in wide ranging subjects that (apart from surgery and medicine) included chemistry, mathematics, languages, poetry, even theology and philosophy.

Around the end of 1797, Humphry Davy's life took a dramatic turn. He chanced upon the English translation of *Traité Élémentaire de Chimie, présenté dans un ordre nouveau, et d'après des découvertes modernes* (*Elements of Chemistry in a New Systematic Order containing all the modern discoveries* by Robert Kerr in 1790) and William Nicholson's *Dictionary of Chemistry (Volumes 1 and 2)* and decided to choose chemistry as his career option. He also became acquainted with Gregory Watt, son of James Watt when he was a boarder at his mother's boarding house, Davies Giddy and later Davies Gilbert – all successful men of science. Just as Mr. Tonkin and Mr. Dunkin had shaped his early youth, these three men, each in his own unique manner, were instrumental in setting Humphry on his career as an experimental chemist. At this time (1798), Dr. Thomas Beddoes was interested in studying the medical properties of a few selected gases including the recently discovered nitrous oxide and established a research institute called *"Pneumatic Institution for Relieving Diseases by Medical Airs"* at Bristol. Gilbert recommended Humphry Davy for a position as a superintendent at Dr. Beddoes's institute. Aided by Mr. Tonkins intervention, Humphry Davy was released from his apprenticeship with Dr. Borlase. Aged just 19 years, he took up the position with Beddoes that launched him on his scientific journey.

Becoming an experimental chemist

Davy's career as a chemist had an inauspicious beginning. In early 1799, Beddoes published in his West Country Contributions a collection of papers that included two of Davy's papers - *"On Heat, Light and the Combinations of Light; with a new Theory of Respiration and Observations on the Chemistry of Life"* (in which he argued against Lavoisier's view of caloric (heat) stating heat was motion but light was matter) and "*On the Generation of Phos-oxygen (Oxygen gas) and the Causes of the Colours of Organic Beings*" - based on early experiments he had done as an adolescent when he was an apprentice in Penzance. While these experiments were designed intelligently, they suffered from serious drawbacks. They were poorly conducted and his conclusions instead of being based on empirical findings, were conjectures based on his imagination. For example, in his paper on heat and light,

Davy tried to prove experimentally that as heat could be generated by friction when two pieces of ice were rubbed, heat was really not an element or material as claimed by Lavoisier and that "light was a matter of a peculiar kind". *"I am now as much convinced of the non-existence of caloric as I am of the existence of light"* (Davy's letter to Gilbert Davies, 22 February 1799, From Hunt [9]). In his paper on the generation of phosoxygen, Davy tried to prove that oxygen was a compound and hence should be called "phosoxygen". He went to the extent of categorically stating that the phosoxygen theory explained the blue color of the sky, electricity, red color in roses, the aurora borealis, melanin pigmentation in people from Africa, the fire of falling stars, thought, perception, happiness, and why women are fairer than men (Humphry Davy "*On the Generation of Phosoxygen (Oxygen gas) and the Causes of the Colours of Organic Beings*") (From Kenyon [10]). His hasty conclusions did not find acceptance and his theory was dismissed. Davy was humiliated by this rejection and bitterly regretted publishing his hasty generalizations which had no real basis. Much later, Davy regretted publishing his immature ideas and termed his hypotheses as *"the dreams of misemployed genius which the light of experiment and observation has never conducted to truth"* (From Hunt [9]). Chastened by this experience, Davy refined his experiments and in the process, became one of the greatest experimental chemists.

"I thank God that I was not made a dexterous manipulator, for the most important of my discoveries have been suggested to me by my failures" - Humphry Davy (From Mellor [11]).

Inhaling nitrous oxide

In Beddoes' institute, 20 year old Davy became well versed in chemistry and carried out experiments on the medicinal benefits or ill effects of inhaling various gases. Davy was encouraged to do research on the effects of nitrous oxide. At that time it was believed that nitrous oxide gas was toxic and was the cause of many diseases as it had a bad effect on animal tissues. (Dr. Mitchell, a little known doctor held the view that nitrous oxide spread contagious diseases). Young Davy decided to test the veracity of this belief on himself. As part of the experiment, he recklessly inhaled nitrous oxide gas several times a day (some days inhaling as much as sixteen quarts of pure nitrous oxide for close to seven minutes at a time). Far from causing any damage to his tissues, the only effect inhalation of nitrous oxide had on him was one of a relaxed and happy feeling of intoxication.

26th December 1799 was an important day in the life of Humphry Davy. It marked his final experiment on the effect of nitrous oxide. The experiment took place in the laboratory of the Pneumatic Institution under the supervision of the physician Dr. Robert Kinglake. Stripped to his waist, he went inside the special sealed chamber, built by his friend James Watt, for the experiment of inhaling nitrous oxide. At intervals of five minutes, Dr. Robert Kinglake released twenty quarts of nitrous oxide till Davy became unconscious. Davy described the effects of that experiment in his book *Researches, Chemical and Philosophical, chiefly concerning Nitrous Oxide* (1800), *"...The moment after I began to respire 20 quarts of unmingled nitrous oxide, a thrilling, extending from the chest to the extremities, was almost immediately produced. I felt a sense of tangible extension highly pleasurable in every limb; my visible impressions were dazzling, and apparently magnified, I heard distinctly every sound in the room and was perfectly aware of my situation. By degrees, as the pleasurable sensations increased, I lost all connection with external things; .... I existed in a world of newly connected and newly modified ideas. I theorised—I imagined that I made discoveries. When I was awakened from this semi-delirious trance by Dr. Kinglake, who took the bag from my mouth, indignation and pride were the first feelings produced by the sight of the persons about me. My emotions were enthusiastic and sublime; and for a minute I walked round the room, perfectly regardless of what was said to me. As I recovered my former state of mind, I felt an inclination to communicate the discoveries I had made during the experiment. I endeavoured to recall the ideas, they were feeble and indistinct; one collection of terms, however, presented itself: and with the most intense belief and prophetic manner, I exclaimed to Dr. Kinglake, Nothing exists but thoughts!—the universe is composed of impressions, ideas, pleasures and pains!"* (From Davy [12]).

Based on his personal experience, Davy announced that it was perfectly safe to inhale nitrous oxide. This announcement immediately brought both Davy and his institute to public notice. Inhaling nitrous oxide gas in social gatherings of the high society became extremely fashionable in London for its intoxicating effect. Poets Coleridge and Southey were among those who indulged in this pleasurable activity and became Davy's ardent admirers. Davy also experienced that inhaling nitrous gas deadened pain and casually mentioned that it could be used in surgery to relieve pain. He remarked *"as nitrous oxide ... appears capable of destroying physical pain, it may probably be used with advantage during surgical operations in which no great effusion of blood takes place"* (From Davy [12]). He did not pursue with clinical trials to find out the feasibility of the use of nitrous oxide as an anesthetic during surgery. It took more than forty years before nitrous oxide was tried as an anesthetic.

After the initial excitement, Davy got totally absorbed in studying the properties of nitrous oxide in greater detail. Based on the findings from his studies, he published the treatise mentioned earlier and gained recognition as an excellent chemist. Davy continued his experiments on the three oxides of nitrogen, namely NO_2, NO, and N_2O. The analytical results of his experiments substantiates one of the tenets of Dalton's atomic theory which Davy had dismissed as *"rather more ingenious than important" (When two elements combine and form more than one compound, the masses of one element that react with a fixed mass of the other are in the ratio of small whole numbers).* It was years later that he acknowledged the importance of Dalton's work.

Around 1799, the Royal Institution was established by distinguished scientists for "diffusing knowledge and facilitating the general introduction of useful mechanical inventions and improvements; and for teaching, by courses of philosophical lectures and experiments, the application of science to the common purposes of life". The Royal Institution was looking for a lecturer in chemistry who could successfully implement its charter. Davy got a big break when Professor Hope of the University of Edinburgh who had been impressed by Davy's work at the Pneumatic Institute strongly recommended Davy to Count Rumford (one of the founders of the Royal

Institution) for the job. Humphry Davy was interviewed for the job in 1801 February, by a committee consisting of Joseph Banks, Benjamin Thompson and Henry Cavendish and it recommended *"that Humphry Davy be engaged in the service of the Royal Institution in the capacity of assistant lecturer in chemistry, director of the chemical laboratory, and assistant editor of the journals of the institution, and that he be allowed to occupy a room in the house, and be furnished with coals and candles, and that he be paid a salary of 100l per annum"* (From Paris [6], Hunt [9]).

> The self-educated chemist from the countryside with no sophistication arrived in London in March 1801 to join the Royal Institution as an assistant lecturer in chemistry. He did not have to look back.

Lectures and Research at the Royal Institution

Davy was a brilliant lecturer. He selected galvanism or electricity as the topic of his maiden public lecture that he delivered at the Royal Institution on 25th April 1801. During the course of his lecture, Davy demonstrated the power of galvanism or electricity produced by chemical means to cause movement in the amputated legs of frogs and to catalyse the isolation

(From "Humphry Davy", Wikipedia)

A young Humphry Davy gleefully works the bellows in this caricature by James Gillray of experiments with laughing gas at the Royal Institution. Dr Garnett is the lecturer, holding the victim's nose.

of metals from aqueous acids. The lecture was a tremendous success. Davy's flair in delivering the lecture, the theatrical effects, coupled with his clear understanding of scientific advances, brought him instant fame as a public lecturer.

By the time he gave his final lecture of the series in June, the audience had grown manifold and according to Davy himself *"There was Respiration, Nitrous Oxide, and unbounded Applause. Amen!"* (From "Humphry Davy" Wikipedia [13], Fullmer[5]). This convinced the authorities of the Royal Institution of his abilities and within a short period at the age of 23 years, he was made a professor of chemistry. As professor of chemistry and director, he transformed the Royal Institution into a premier institution for the promotion of science and a centre for research.

Foray into electrochemistry: Davy had been excited by the experiments in electricity announced by Alessandro Volta of Italy using the "electric pile". Davy at oncc realized that a chemical reaction taking place in the electric pile was responsible for producing electricity and decided to focus his research in this area.

Nicholson and Carlisle had decomposed water into hydrogen and oxygen for the first time using Volta's pile and they also decomposed many salts in their aqueous state. Davy had worked on Galvani's research even before he joined the Royal Institution but he could not pursue his research better then as he did not have a large battery available to conduct his experiments relating to electrochemistry. Davy's position as the professor of chemistry gave him the perfect platform to launch his experiments in electrochemistry. While the earlier

A voltaic pile

(From "Humphry Davy", Wikipedia)

electric piles had consisted of two different metal plates or one metal plate and a charcoal plate with a fluid between the plates, Davy's apparatus consisted of a single metal plate and two fluids one of which could oxidize the surface of the metal. Davy tried various combinations of metal and liquids. He soon discovered that when an electrical current was passed through certain substances, they decomposed into their constituent elements. Davy also realized that the voltaic pile and electrolysis worked in a similar fashion.

One of the major tasks he had to undertake on joining the Royal Institution was to give lectures on the chemical processes involved in tanning. Davy conducted a number of experiments to understand the chemistry of tanning and delivered a lecture on his findings at the Royal Society in 1803. In 1802, the board of agriculture made a request to Davy to give a course of lectures on agricultural chemistry. The Royal Institution gave him permission to undertake the task and he gave the lectures in 1803. These lectures were so successful that he gave them for ten consecutive years and later published them as a book titled *Elements of Agricultural Chemistry* (1813).

> Despite the obvious disadvantages of his background, self-taught education and pursuit of social acceptance, Davy was highly respected at the Royal Society for his professional brilliance and extraordinary ability as a lecturer. He was elected fellow of the Royal Society in 1803 at the age of 25 and as one of the two secretaries the same year. He went on to become its president from 1820 to1827.

Davy was an entertainer par excellence. Towards the end of 1810, public lectures on science had become a source of entertainment to both the elite and commoners of London. Davy was a master at creating an atmosphere of sheer magic. With his understanding of science, breath-taking demonstrations, theatrical presentation and sheer oratory, he mesmerized his audience. In 1808, he had the enormous 600-plate voltaic battery and molten fluids in flasks brought to the lecture hall of the Royal Institution to demonstrate live how electrochemistry works. The audience was spell-bound when as if by magic, the process of electroplating of one electrode by a metal and bubbles of oxygen coming out from the other electrode began when he connected the huge voltaic battery to the electrolytic cell, The August 1808 issue of *Monthly Magazine* published an engraving spread over two pages of "Professor Davy's great Galvanic Apparatus at the Royal Institution, by which he has effected the decomposition of the Alkalies."

Young Faraday who was present in one of Davy's lectures had this to say about the effect of Davy's lecture on the audience: "*Sir Humphry Davy proceeded to make a few observations on the connections of science with other parts of polished and social life. Here it would be impossible for me to follow him. I should merely ... destroy the beautiful and sublime observations that fell from his lips. He spoke in the most energetic and luminous manner of the Advancement of the Arts and Sciences. Of the connection that had always existed between them and other parts of a Nation's economy. During the whole of these observations his delivery was easy, his diction elegant, his tone good and his sentiments sublime*" (From Timmons [14]).

Humphry Davy and Michael Faraday

"Davy's greatest discovery was Michael Faraday" (From Harvey [15]).

Michael Faraday most unexpectedly got the opportunity to work at the Royal Institution under Sir Humphry Davy when Davy damaged his sight by a serious accident in his laboratory while he was working with nitrogen trichloride. Davy hired Faraday as a bottle washer to maintain the glassware and to keep all the apparatus ready for him to conduct experiments. In practice, Faraday was his assistant in the laboratory. This intellectual apprenticeship proved crucial for the scientific world. Davy shared a close professional as well as personal relationship with Faraday.

Davy and Faraday came from remarkably similar backgrounds. Their families were poor, both had to be apprenticed to take care of the family responsibilities and both were self educated, driven by a desire to do science. Both were gifted with imagination but with a difference. While Davy's was poetic, Faraday could visualise complex concepts. Both were outstanding public speakers but with a difference. Davy was a natural orator. Faraday had to overcome a childhood speech impediment by sheer determination to become an outstanding speaker. His Christmas lectures at the Royal Institution which were attended by Royalty as well as by common people are classic examples. Yet they were completely different as persons and in their approach to life. While Davy sought recognition and acceptance by the elite society of London and Royalty (and even married an aristocratic lady), Faraday was deeply religious, worked alone in his laboratory with his assistant, wrote his papers and refused all honours including the Presidentship of the Royal Society and a knighthood, preferring to be just 'Mr. Faraday'.

Bakerian lectures

The Bakerian Lecture of the Royal Society was established in 1775 by the Royal Society from a gift of 100 pounds by Henry Baker. This prestigious lecture in physical sciences is given annually by a Fellow of the Royal Society. Initially, the lecturer received an honorarium of 10 pounds. Today, the honorarium is 1000 pounds! Davy was invited by the Royal Society to deliver the prestigious Bakerian lectures in 1806. He gave six Bakerian lectures at the Royal Society – five lectures between 1806 and 1810 and the final one just four years before his death. Davy's Bakerian lectures were outstanding for the theatrical flourish with which he presented his discoveries.

In his first Bakerian lecture *'On some Chemical Agencies of Electricity'* delivered at the Royal Society in 1806, Davy presented the results of his experiments thus: *'Hydrogen, the alkaline substances, the metals and certain metallic oxides are attracted by negatively electrified metallic surfaces, and repelled by positively electrified metallic surfaces; and conversely acid and substances such as oxygen are attracted by positively electrified metallic surfaces and repelled by negatively electrified metallic surfaces; and these attractive and repulsive forces are sufficiently energetic to destroy or suspend the usual operation of elective affinity'* (From Davy [16]). Davy's contemporary and one of the great Swedish chemists, Jacob Berzelius, called the lecture one of the most outstanding *'memoirs in the history of chemical theory'*. Davy also outlined the theory of chemical affinity based on the evidence obtained from his experiments and in conclusion he suggested that "*the electric decomposition of neutral salts might in some cases....... lead to the isolation of the true elements of bodies*" (From Davy [16(a)]).

Davy was awarded the annual medal established by Napoleon by the French Institute for the best experiment conducted each year on galvanism, even though the two countries were at war. Davy remarked to Thomas Poole on why he accepted the award *"But if the two countries or governments are at war, the men of science are not. ..... we should rather, through the instrumentality of the men of science soften the asperities of national hostility"* (From Beer [17]).

Davy delivered his second Bakerian lecture in 1807 on the preparation of potassium and sodium by electrolytic process using his battery.

"When [Humphry Davy] saw the minute globules of potassium burst through the crust of potash, and take fire as they entered the atmosphere, he could not contain his joy—he actually bounded about the room in ecstatic delight; some little time was required for him to compose himself to continue the experiment" (From Davy [18]). Samuel Taylor Coleridge who attended almost all of Davy's lectures *"to renew my stock of metaphors"* (From Hartley [19]) wrote about this lecture to Dorothy Wordsworth, sister of poet William Wordsworth, *"In November 1807 Davy gave his famous Second Bakerian Lecture at the Royal Society, in which he used Voltaic batteries to 'decompose, isolate and name' several new chemical elements, notably sodium and potassium"* (From Griggs [20]).

Three days short of his thirtieth birthday, Davy gave his third Bakerian lecture in the main hall of the Royal Society in 1808 in which he gave a detailed description of the preparation of boron. This was one of the longest lectures but fell short of his first two lectures in its importance. Assuming boron to be a metal, Davy initially named it boracium. Based on empirical evidence obtained from his experiments, Davy disproved Gay-Lussac's view that potassium was not an element but a compound containing hydrogen. Ironically, he did not believe for a while that carbon, phosphorous and sulfur were elements (though he corrected his view later). Davy delivered his fourth Bakerian lecture in November 1809 when he presented more experimental proof of potassium being an element, and the fifth lecture in 1810 in which he stated that oxymuriatic acid was not a compound but an element and suggested that it be called 'Chlorine'.

"Every discovery opens a new field for investigation of facts, shows us the imperfection of our theories. It has justly been said that the greater the circle of light, the greater the boundary of darkness by which it is surrounded" (From Davy [21]).

Davy's outstanding achievements

Davy made outstanding contributions to many areas:

- Discovery of the medical properties of nitrous oxide or laughing gas (1799-1800).
- Providing proof that it was not using two different metals that made the Voltaic Pile work. Davy showed that it was possible to build an electric cell with one metal plate and two fluids, one of which could oxidize the surface of the metal.
- Proving that alkalis and alkaline earths were oxides (compounds formed by oxygen combining with the metals). This challenged Lavoisier's assertion that oxygen was an integral part of an acid.
- By an ingenious but simple experiment (when two big ice cubes or blocks were rubbed together, they melted without any addition of external heat) disproving the firmly held belief of Lavoisier and others that caloric (heat) was an element.
- Performing the first electrochemical decompositions and isolating the elements potassium and sodium (1807).

- Devising a method for separating the two alkali elements - sodium and potassium. He observed that while potassium perchlorate was insoluble in 97% alcohol, sodium perchlorate was soluble in it. Using this difference, he developed the method of separating the two elements.
- Isolation of pure barium, strontium, calcium, and magnesium (1808) by electrolysis. Other scientists had done a lot of spadework in this field. For example, Scheele had discovered baryta (barium oxide), Berzelius and Pontin had electrolyzed lime in mercury and obtained calcium amalgam. But Davy succeeded in isolating the metals. Although boron compounds were known for centuries, the element boron was discovered only in 1808 by Humphry Davy, Gay-Lussac and Thenard. Davy prepared boron by making use of the reducing property of potassium.
- Inventing the first arc lamp by connecting two wires to an external battery and then attaching a charcoal strip between the other two ends of the wires. This arrangement charged the carbon strip which started to glow.

Chlorine was one of Davy's important areas of research. Many have compared the importance of his research on chlorine to his work on alkali metals. Scheele of Sweden was the first to isolate chlorine but failed to recognise the green gas with its distinct pungent smell as an element. He called it "dephlogisticated marine acid". Davy conducted many experiments reacting "dephlogisticated marine acid" with "oxymuric acid" (as ammonia was called at that time) to detect the presence of oxygen as Lavoisier considered chlorine was an oxide of a hitherto unknown radical. Davy found that his experiments yielded only muriatic acid (hydrochloric acid) and nitrogen. He then attempted to remove oxygen from the gas by exposing it to white hot carbon. In spite of conducting the experiment repeatedly, Davy was unable to get any oxygen containing compound or oxygen gas. This convinced him that the gas was an element and he named it chlorine, the Greek word "chloros" meaning yellow green. The series of experiments he conducted in 1810 with muriatic acid (HCl) and mercury which yielded calomel and hydrogen are equally important.

Davy was not the first chemist to prepare iodine. It was discovered for the first time in 1811 by Bernard Courtois of France. Courtois was trying to obtain potassium nitrate in the laboratory from burning seaweed kelp and accidentally discovered that when excess sulfuric acid was added to the ash and heated, it gave out a pungent-smelling violet-coloured vapour. On coming in contact with cold surfaces, the vapours condensed and formed almost black crystals. Unfortunately, Courtois did not realize its importance. As he could not pursue the investigation further, he gave some of the scrapings from the vessel to Joseph Gay-Lussac and to Humphry Davy who was in Paris at that time. Both Gay-Lussac and Davy (working in his portable lab in his hotel room) established that the black crystals prepared by Bernard Courtois were crystals of a new element belonging to the same family as chlorine. It was called iodine, after the Greek word *iodes* meaning violet.

Davy made iodine pentoxide, an important iodine compound, for the first time in 1815. It is a colorless, odorless crystalline substance and has high density. It has strong oxidizing property, and sometimes even forms explosive compounds with certain substances (that can be oxidized).

Davy observed that the oxidation of alcohol vapor in air was induced by platinum

Silica from Silicates: Davy was successful in developing a method to prepare silica from silicates by decomposition. He did this by treating silicate with hot HCl.

$$SiO_4^{4-} + 4H^+ \longrightarrow SiO_2 + 2HOH$$

Discovery of Cool flame: Davy accidentally discovered "cool flame" in the 1810s. He observed that unlike conventional flames, certain types of unusual flames produced by the chemical reaction of certain air-fuel mixture neither burnt his fingers nor ignited a matchstick. Also, at certain temperature and composition it did not require an external source like a hot material or a spark to ignite it! These unusual flames could also be converted to a conventional flame.

Davy, the multifaceted genius

Davy was a born experimenter and a designer of instruments to carry out experimental work. Apart from his outstanding contributions to chemistry, he is credited with making the first electric light. He conducted many experiments with electricity and during the course of one experiment, when he connected the piece of carbon and his battery with a wire, the carbon piece glowed producing light.

> Many of Davy's discoveries had an unexpected spin off – they contributed to mining, tanning and agricultural industries adopting a scientific approach towards production practices which in turn led to improved production.

Apart from being an outstanding researcher, Davy was interested in finding practical solutions to problems faced in various fields. His processes for desalinating sea water and connecting zinc plates to the copper-clad steel of the ship's submerged hull to prevent erosion of the hull are typical examples of this. Charles Babbage said of this particular discovery *"The influence of electricity in producing decompositions..... can hardly be said to have been applied to the practical purposes of life, until the same powerful genius [Davy] which detected the principle, applied it, by a singular felicity of reasoning, to arrest the corrosion of the copper-sheathing of vessels. ... this was regarded by Laplace as the greatest of Sir Humphry Davy's discoveries"* (From Babbage [22]).

Davy lamp

Davy was concerned about the serious problem of frequent explosions caused by the accumulation of an inflammable mixture of methane and air in underground coal mines. He wanted to find a solution to this problem faced by miners. He conducted a number of experiments with the firedamp samples obtained from the coal mines at Newcastle. Davy developed a safety lamp in which he used metal tubes instead of glass tubes. This prevented the ignition of the inflammable gases in the mine and thus prevented explosions in the coal mines. On November 9, 1815, he presented his findings in a paper at the Royal Society in which he showed metallic tubes were better than glass ones for covering the flame as the metallic tubes dissipated the heat. This rendered the temperature of the initial explosions of the outer layers of the gas harmless as they could not ignite the other layers of gas. In 1816, he improved the lamp further by using wire gauze instead of tubes to surround the flame. He succeeded in fabricating a lamp that allowed miners to work safely at great depths in the coal mines. This lamp came to be known as Davy lamp but Davy did not take a patent as he felt strongly about its usefulness to the mining society.

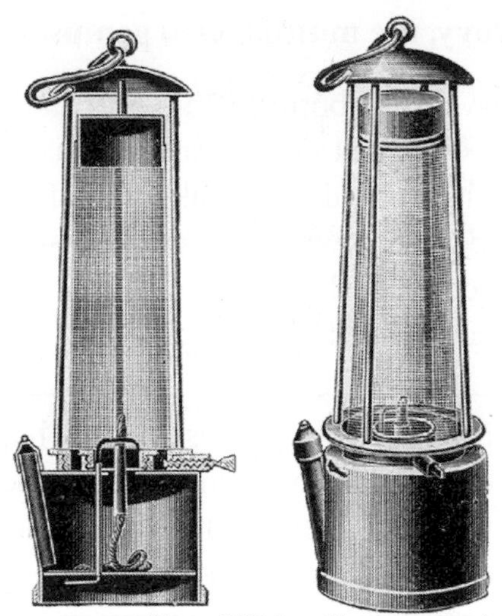

(From Wikimedia commons)

Honours

Davy received many honours for his outstanding contributions to experimental chemistry. Some of them are: Napoleon Prize from the Institut de France (1807), Fellow of the Royal Society (1803), Copley Medal (1805), Rumford Medal (1816) and Royal Medal (1827) of the Royal Society, Knighthood, President of the Royal Society (1820 to 1827).

Davy was made a baronet in 1818 – the ultimate social recognition for which he strived throughout his life. After Sir Isaac Newton, Davy was only the second English scientist to be conferred this unique honour for his service to science. Among the many honours and recognitions he received, Knighthood and baronetcy were the two honors that he cherished most.

"To me there never has been a higher source of honour or distinction than that connected with advances in science. I have not possessed enough of the eagle in my character to make a direct flight to the loftiest altitudes in the social world; and I certainly never endeavored

to reach those heights by using the creeping powers of the reptile, who in ascending, generally chooses the dirtiest path, because it is the easiest" - Humphry Davy (From Davy [23]).

Davy the man

While Davy held strong views on the importance of science, he also understood its limitations to explain all the mysteries of nature. He was convinced that *"Nothing is so fatal to the progress of the human mind as to suppose that our views of science are ultimate, that there are no mysteries in nature, that our triumphs are complete and that there are no new worlds to conquer"* (From Davy [24]). His overriding ambition to achieve fame and acceptability by high society often made him display petty jealousy but he also gave his time and services unstintingly to problems faced by the common man. He had a wry sense of humour.

The ultimate tribute to the genius of Davy was paid by Davies Gilbert, President of the Royal Society in his address to the Royal Society on Davy's death.

"The poetic beauty of Davy's mind never seems to have left him. To that circumstance I would ascribe the distinguishing feature in his character, and in his discoveries,—a vivid imagination sketching out new tracts in regions unexplored, for the judgment to select those leading to the recesses of abstract truth" (From Davy [25]).

Davy showed the same energy and passion in pursuing his many interests, be it testing the result of inhaling gas mixtures, conducting serious experiments to discover new elements or delivering scientific lectures that kept the audience spell-bound, finding solutions to practical problems or merely writing poetry. Davy threw himself wholeheartedly into the task at hand. Many of the literary giants of the day like Coleridge and Southey were his personal friends and attended his public scientific lectures to learn something new all the time. Coleridge was convinced that *"had (Davy) not been the first chemist, he would have been the first poet of his age"*. He went to hear Davy "*to increase his stock of metaphors."*

Southey declared that *"he had all the elements of a poet; he only wanted the art"* (From "Humphry Davy", *Wikipedia* [26]).

Final years

Coleridge wrote of Davy in 1801 *"chemistry tends...to turn it's [sic] Priests into Sacrifices"* (From Kenyon [10]).

Davy's health became a serious concern as he fell frequently ill even during his most prolific years. By 1823, his health showed a steady decline. In 1827, he suffered a minor paralytic attack. His doctors advised him to go abroad. He returned for a brief period to England. While his failing health forced him to give up the presidency of the Royal Society, it did not prevent him from continuing writing. It was not about chemistry but about fishing. He wrote a book on fishing titled "Salmonia: or Days of Fly Fishing" complete with illustrations from his own drawings. Sir Humphry Davy eventually left England in 1828, never to return. He died of heart failure in Switzerland on May 29, 1829 at an young age of 50. He was mourned by scientists and poets alike.

'Davy's gone. Surely these are men of power, not to be replaced should they disappear, as one alas has done' - William Wordsworth (From Clayden [27]).

References

1. Humphry Davy, '*Consolations in travel, or, The last days of a philosopher*', 1830.
2. Richard Holmes, '*Humphry Davy and the Chemical Moment*', *Clinical Chemistry* November 2011 vol. 57 no. 11, 1625-1631; 2011 The American Association for Clinical Chemistry.
3. Humphry Davy, John Davy (Ed), Discourse Delivered at the Royal Society November 30, 1825. *The Collected Works of Sir Humphry Davy, Discourses delivered before the Royal Society,* Vol VII, 1840.
4. Richard Holmes, *The Age of Wonder: How the Romantic Generation Discovered the Beauty and Terror of Science*, Pantheon Books, 2008.

5. June Z. Fullmer, *Young Humphry Davy:The Making of an Experimental Chemist,* Volume 237, 2000.
6. John Ayrton Paris, *The Life of Sir Humphrey Davy, 1831.*
7. John Ayrton Paris, *The Life of Sir Humphry Davy,* Volume 1, 1831.
8. John Davy (ed*), The Collected Works of Sir Humphry Davy,* vol.1, Memoirs of his life, London, 1839.
9. Robert Hunt, *Dictionary of National Biography*, 1885-1900, Volume 14, Humphry Davy, Wikisource - Davy, Humphry (DNB00) 9 March 2012.
10. T. K. Kenyon, Chemical Heritage Magazine, ***Science and Celebrity***, Winter 2008/9 edition.
11. J. W. Mellor, *Modern Inorganic Chemistry* (1925).
12. Humphry Davy, J. Davy (ed.), *The Collected Works of Sir Humphry Davy, Researches, Chemical and Philosophical,* London, vol. III (1839).
13. *Humphry Davy* in Wikipedia, *The Free Encyclopedia, (Source:* Holmes, Richard, *The Age of Wonder*. (2008). Pantheon Books).
14. Todd Timmons, *Makers of Western Science: The Works and Words of 24 Visionaries from Copernicus to Watson and Crick*, 2012.
15. Paul Harvey (ed.), "Michael Faraday", in *The Oxford Companion to English Literature* (1932), 279.
16. Humphry Davy, John Davy (ed), *The Collected works of Sir Humphry Davy,* Vol V., Bakerian Lectures and Miscellaneous papers from 1806 to 1815. London, 1840. (a) Humphry Davy, Bakerian Lecture, 'On Some Chemical Agencies of Electricity', *Philosophical Transactions of the Royal Society*, 1807, **97**.
17. Quoted in Gavin de Beer, *The Sciences were Never at War* (1960), 204 & p 885, *The New Scientist*, 7 April, 1960, vol.7, No. 177).
18. J. Davy (ed.), Edmund Davy Quoted in *Memoirs of the Life of Sir Humphry Davy*, 1839, p109.
19. Harold Hartley, *Humphry Davy* 1971, 45.
20. Earl Leslie Griggs (ed.), *Collected Letters of Samuel Taylor Coleridge,* (1956), Vol. 3, 38.

21. Humphry Davy, J. Davy (ed.), *The Collected Works of Sir Humphry Davy,* 'Consolations in Travel—Dialogue V—The Chemical Philosopher' (1840), Vol. 9, 362.

22. Charles Babbage, *Reflections on the Decline of Science in England, and on Some of Its Causes*, (1830), 16.

23. Humphry Davy, John Davy (ed.), *The Collected Works of Sir Humphry Davy*, Vol. IX, *Salmonia and Consolation in Travel*, London, 1840.

24. Humphry Davy, John Davy (ed.), *The Collected Works of Sir Humphry Davy,* Vol. VIII, *Agricultural Lectures Part II and other lectures,* London, 1840.

25. Presidential Address to the Royal Society on Davy's Death, 1829. Quoted in J. Davy, *Fragmentary Remains of Sir Humphry Davy* (1858).

26. Humphry Davy, Wikipedia, *The Free encyclopedia, Retrieved Feb 18, 2015*

27. P. W. Clayden, *Rogers and his Contemporaries*, London, Vol.2, 1889.

Wellcome Library, London, M0004638 Portrait of Sir Humphry Davy, 1st Baronet, FRS (1778 – 1829), Creative Commons Attribution only license CC by 4.0 http://creativecommons.org/licenses/by/4.0/

"New Discoveries in Pneumatics". Humphry Davy (1778-1829) lecturing at the Royal Institution. Print by James Gillray, 1802, (Source=[http://www.chemsoc.org/networks/learnnet/periodictable/scientists/davy.htm]) (Humphry Davy, Wikipedia).

Alessandro Volta's electric battery, Author: GuidoB Attribution: I, GuidoB (From Wikipedia, The free encyclopedia).

Bibliothek allgemeinen und praktischen Wissens für Militäranwärter Band III, 1905 / Deutsches Verlaghaus Bong & Co Berlin * Leipzig * Wien * Stuttgart - Scan by Kogo (Wikimedia commons).

4. JONS JACOB BERZELIUS (1779–1848)

Swedish pioneer who wrote the first chemistry text book

"Those who are unacquainted with the details of scientific investigation have no idea of the amount of labour expended in the determination of those numbers on which important calculations or inferences depend. They have no idea of the patience shown by (a) Berzelius in determining atomic weights" (From Tyndall [1]).

Early years

Berzelius was born on August 20, 1779 in Väversunda, Östergötland, Sweden, into an educated family. His father Samuel Berzelius was a teacher in a gymnasium, and mother, Elizabeth Dorothea, was a homemaker. Young Berzelius had an emotionally difficult childhood as his father died when he was only four years old and his mother remarried the pastor of Norrkoping, himself a father of five children. Unfortunately tragedy struck Berzelius again five years later when his mother died, and his stepfather remarried when he was 11 years old. Berzelius and his sister were sent to the care of Magnus Sjösteen, their maternal uncle. There was constant friction between the Berzelius siblings and his seven cousins. This frequent

(From "Jons Jacob Berzelius", *Wikipedia*)

upheaval in his young life affected Berzelius. He joined Linköping gymnasium in 1793 at the age of 14 years, as much to escape the constant fights with his cousins as to pursue studies. After a year at the gymnasium, Berzelius took the job of a tutor on a nearby farm to pay for his education. Even though his work at the farm was tough and living conditions were harsh (he had to share his sleeping quarters with sacks of potatoes stored there), he developed an abiding love for nature. This experience made him pursue natural science instead of becoming a clergyman. Fortunately, he got a scholarship which was sufficient to resume his studies. Later he opted for a medical degree.

Berzelius joined Uppsala University in 1796 to study medicine at the age of seventeen. Uppsala University was well known for its tradition in chemistry. Ekeberg who discovered tantalum taught there and Scheele had discovered oxygen there. Berzelius was excited by Scheele's experiment and decided to try it in the closet of his room and to his delight succeeded in preparing oxygen. *"I have seldom experienced a moment of such pure delight as when the glowing stick held in the gas ignited in it and illuminated my windowless laboratory with unusual brightness"* - *Berzelius* (From Jorpes & Steele [2]). It is humorously remarked that this flame decided Berzelius' choice of chemistry as his preferred subject.

In the summer of 1800, he was once again in dire financial need and decided to take a year off by staying with his aunt. Here, Berzelius had an unexpected stroke of luck when his uncle first sent him to work as an assistant in a pharmacy (where Berzelius learnt glass-blowing) and later as an assistant to the chief physician of Medevi mineral springs. His job there was to analyse the spring water for its mineral content. This set Berzelius firmly on the path of a scientific career. Around this time, news of Alessandro Volta's electric pile created a stir in scientific circles of Europe. The electric pile as a reliable source of continuous electric current was demonstrated in Uppsala. Berzelius at once saw how it could be used to study the effect of galvanic current on diseases, the topic of his thesis. He decided to build his own electric pile by using sixty zinc disks and alternating them with sixty highly polished copper coins (silver would have been too expensive). While working in Medevi, Berzelius studied the effect of galvanic current on a number of diseases. His experiments indicated no beneficial effect of galvanic current on diseases but resulted in developing an abiding interest in electrochemistry. He obtained the MD degree in 1802 and his mentor (Sven Hedin) obtained a job for him as an unpaid assistant at the college of medicine so that he could pursue his studies in chemistry. He was lucky to be a boarder in the house of Hisinger, who owned mines and was interested in chemistry and mineralogy. Hisinger was also an important member of the Galvanic society of Sweden which owned the largest voltaic pile in Sweden. The electrochemical experiments they conducted were the forerunners of Berzelius' future work in this field. In 1803, Berzelius showed that his electrochemical cell had enough power to decompose chemical compounds into their constituents as pairs with opposite electric charges.

Berzelius published his first book, "Afhandling om Galvanismen" (Treatise on Galvanism), a review of the studies conducted on the impact of electricity on various minerals and salts in 1802. His collaboration with Hisinger proved to be an advantage as Berzelius got permission to use the voltaic pile of the Galvanic society. In 1803, using the large voltaic pile, Berzelius and Hisinger conducted experiments on various salts of sodium, calcium, potassium etc. and discovered that electric current decomposed all the salts with oxygen, oxidized compounds as well as acids collecting at the positive pole and alkalies and combustible substances collecting at the negative pole. The exception to this general rule was when an acid acquired a lower oxidation number, it was accumulated at the negative pole. Thus, the lower oxidation state represented a "combustible body." Similar results were obtained and extended by Humphry Davy who isolated sodium and potassium during 1806 - 1807. The results from the electrochemical experiments convinced Berzelius of the importance of electric current not only in decomposing chemical compounds but also in binding chemical elements. It also convinced him that an essential constituent of all acids and bases was oxygen. These ideas formed the basis of his theory of electrochemical dualism.

Berzelius was appointed a professor at the Medical College in Stockholm (which was later renamed Karolinska Institute) in 1807 and was elected to the Royal Swedish Academy of Sciences in 1808.

Between 1807 and 1818 Berzelius travelled extensively in Europe visiting all the major centres of scientific research. He met Hans Christian Oersted in 1807. The high point of his travels was his meeting Humphry Davy when he visited England in 1812. Berzelius spent a year in Paris (1818 - 1819) where he worked with Berthollet. He returned to Stockholm via Germany where he created a strong impression on young German chemists (amongst whom Wohler became a personal friend). Through his travels, Berzelius developed strong contacts with most of the leading chemists of his time.

> Berzelius was keen to meet Humphry Davy as their electrochemical studies were very similar. Initially Davy and Berzelius got on well (refer to Berzelius' comment on Davy's Bakerian lecture in Chapter 3), but later the relationship became cool when Davy indirectly heard of Berzelius' negative comments about one of his books. In spite of their common interests, the initial warmth of their relationship was never recovered.

Textbook of Chemistry and atomic weights

***Lärbok i Kemien* (Textbook of Chemistry), the game changer:** *"The devil may write chemical text books ......because every few years the whole thing changes"* – Berzelius (From Jaffe [3]).

Electrochemical experiments

Berzelius was interested in understanding how elements combined to form compounds and was not satisfied with the information available at that time. He therefore, decided to write a chemistry text book (*Lärbok i Kemien*) in Swedish. In preparation for writing *Lärbok i Kemien,* Berzelius conducted a number of experiments on a variety of inorganic compounds to analyse how and in what proportions the constituent elements had combined. Richter of Germany (who called this field of study, stoichiometry in 1792), Joseph-Louis Proust of France and John Dalton of England earlier had proposed principles of chemical combination, but were unsuccessful in providing supporting data. Berzelius meticulously built the apparatus for his experiments and prepared the necessary reagents. He was able to establish empirically that elements combined in fixed proportions by weight in inorganic compounds and demonstrated that this observation conformed to the law of multiple proportions of Dalton. According to this hypothesis, the constituent atoms in inorganic compounds always combined in multiples of whole numbers. Berzelius also discovered that atomic weights were unrelated to the weight of hydrogen. By this discovery, Berzelius was able to establish that elements are not made up of varying numbers of hydrogen atoms (contrary to Proust's hypothesis that all elements were made up of varying numbers of hydrogen atoms). Berzelius analysed innumerable compounds and in the process, discovered elements such as cerium (1803), selenium, and thorium (1828). His students were successful in discovering vanadium (first discovered by Andrés Manuel del Río and later by Nils Gabriel Sefström), lithium (discovered by Johan August Arfwedson) as well as some rare earth elements. Berzelius published the results of his experimental studies in a series of papers. Among them, *Essai sur la théorie des proportions chimiques et sur l'influence chimique de l'électricité* (1819; "Essay on the Theory of Chemical Proportions and on the Chemical Influence of Electricity") is famous. Interpreting the vast experimental data he had accumulated, he was able to establish the atomic weights of many elements. He also determined the formulae of inorganic compounds such as sulfides and oxides.

Discovery of Cerium: His research collaborator Hisinger was instrumental in the discovery of cerium. Hisinger was a keen mineralogist since his boyhood days. He had a collection of minerals and had analysed a number of them. He persuaded Berzelius to analyse a particularly heavy mineral. In 1803, after analyzing the ore, they discovered a hitherto unknown element in it, which they called cerium. Cerium was discovered by Klaproth in Berlin almost simultaneously.

In 1826, Berzelius published a table of atomic weights of elements. While the values of a majority of elements in his table concur with the present day values, a few values differ widely. In determining the atomic weights of 43 elements, his reference was the weight of oxygen for which he assigned 100 as the atomic weight. In addition, there was some confusion how atoms were different from molecules.

Contribution to chemical nomenclature

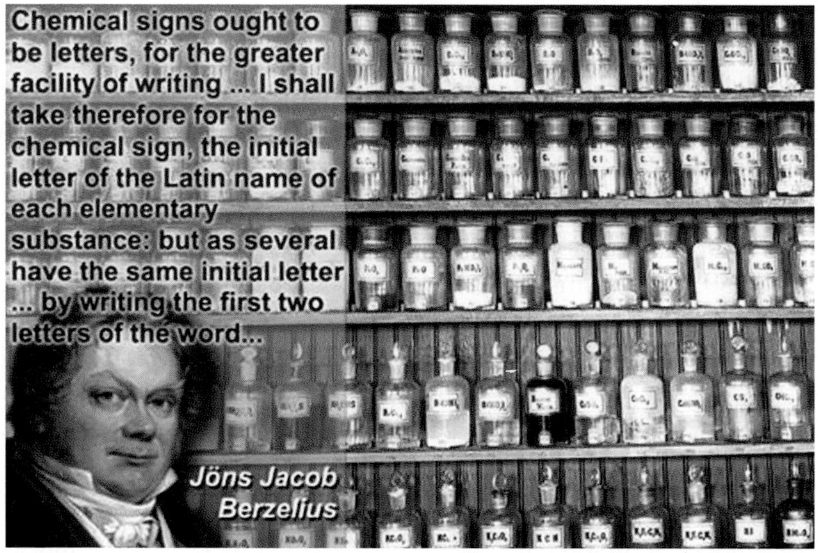

(From www.linkopingshistoria.se)

Chemicals on the shelf in the chemistry laboratory at Miami University (1911)

Nomenclature

Involvement of Berzelius in determining and specifying known as well as newly discovered elements by their atomic weights required long and extensive work. Berzelius felt the need for a new system of chemical nomenclature to make nomenclature easier by a system where the elements were assigned simple and specific written notations which could indicate accurately the composition of any compound (oxygen – O, hydrogen – H, Fe –iron). With this objective, he embarked on a project to create a template for naming the compounds with chemical symbols. He discarded the French names arrived at by Lavoisier and his colleagues and their Swedish translations used by Pehr Afzelius and Anders Gustav Ekeberg.

Instead, Berzelius decided to use Latin names which could be translated into different languages. He succeeded in creating a simple and easy system for naming chemical substances and chemical compounds.

Berzelius on chemical nomenclature

"Chemical signs ought to be letters, for the greater facility of writing, and not to disfigure a printed book ... I shall take therefore for the chemical sign, the initial letter of the Latin name of each elementary substance: but as several have the same initial letter, I shall distinguish them in the following manner:- 1. In the class which I shall call metalloids, I shall employ the initial letter only, even when this letter is common to the metalloid and to some metal. 2. In the class of metals, I shall distinguish those that have the same initials with another metal, or a metalloid, by writing the first two letters of the word. 3. If the first two letters be common to two metals, I shall, in that case, add to the initial letter the first consonant which they have not in common: for example, S = sulphur, Si = silicium, St = stibium (antimony), Sn = stannum (tin), C = carbonicum, Co = cobaltum (cobalt), Cu = cuprum (copper), O = oxygen, Os = osmium, &c" (From H. M. Leicester & H. S. Klickstein [4]).

Berzelius introduced the concept of atomic symbols in 1814 but it took many years for it to be accepted internationally by chemists and his template became the accepted chemical shorthand. His template gave the proportions in which the constituents combined to form a chemical compound. The number of atoms of substance present was denoted by a superscript numerical for example NO^2. His system is used even today with minor changes. The proportions of atoms combining with each other in a compound is now indicated by a subscript instead. For example NO_2 instead of NO^2. In 1819, his expanded system of chemical notation and nomenclature was published in "*Essai sur le théorie des proportions chimiques*" (From Leicester & Klickstein [4]). In this essay, Berzelius presented a comprehensive and thorough system where the theory was arrived at by analyzing the data based on precise experiments.

Berzelius was keenly interested in minerology and contributed to the classification of minerals by their chemical composition instead of their crystalline type as was done till now.

"*As mineralogy constitutes a part of chemistry, it is clear that this arrangement [of minerals] must derive its principles from chemistry. The most perfect mode of arrangement would certainly be to allow bodies to follow each other according to the order of their electro-chemical properties, from the most electro-negative, oxygen, to the most electro-positive, potassium; and to place every compound body according to its most electro-positive ingredient*" – Jöns Jacob Berzelius (From Black [5]).

Electrochemical dualism

Berzelius generalized his findings on the effect of electric current on chemical compounds by stating that all chemical compounds were made up of an electropositive element and an electronegative element, and that they could be identified by their specific constituents with opposite electrical charges. A chemical compound was the result of two atoms carrying opposite charges coming together and neutralizing their charges. Berzelius became famous for enunciating this theory of electrochemical dualism.

"*... every chemical combination is wholly and solely dependent on two opposing forces, positive and negative electricity, and every chemical compound must be composed of two parts combined by the agency of their electrochemical reaction, since there is no third force. Hence it follows that every compound body, whatever the number of its constituents, can be divided into two parts, one of which is positively and the other negatively electrical*" - Jöns Jacob Berzelius (From Encyclopedia.com [6]).

Isomerism

The contributions of Lavoisier to chemistry set in motion the discovery of many new compounds. Chemists assumed that each new compound had its own distinct composition. Analytical studies of some of the compounds such as fulminates and cyanates (studies by Liebig and Wohler) showed that a number of compounds had the same chemical composition. Berzelius became interested in these strange results and decided to look into this phenomenon. He found the same phenomenon in some other compounds. In 1831, Berzelius coined the new term isomerism to describe it. He said *"since it is necessary for specific ideas to have definite and consequently as far as possible selected terms, I have proposed to call substances of similar composition and dissimilar properties isomeric, from the Greek word (composed of equal parts)"* (From Leicester & Klickstein [4]). The German chemist Mitscherlich had used the term isomorphism to describe the phenomenon of different compounds sharing the same crystal structures. Berzelius felt that isomerism best described the phenomenon of some substances having similar composition but dissimilar properties. He also suggested the term allotropy to describe the existence of different forms of the same element (e.g. diamond and graphite as allotropic forms of carbon).

Catalysis

Berzelius was the first chemist to recognize the phenomenon of catalysis and use the term to describe it. Berzelius observed that many chemical reactions took place only when a substance which did not take part in the reaction was present. As he was not sure of the role of this substance, he remarked in 1835 *"This new force, which was unknown until now, is common to organic and inorganic nature. I do not believe that this is a force entirely independent of the electrochemical affinities of matter; I believe, on the contrary, that it is only a new manifestation, but since we cannot see their connection and mutual dependence, it will be easier to designate it by a separate name. I will call this force catalytic force. Similarly, I will call the decomposition of bodies by this force catalysis, as one designates the decomposition of bodies by chemical affinity analysis"* (From Leicester & Klickstein [4]). Much later, Ostwald got the Nobel Prize in Chemistry for his work on catalysis.

Berzelius also coined the term *protein* derived from the Greek word *proteios* meaning primitive in recognition of the importance of these compounds.

Honours

Berzelius was elected Permanent Secretary of the Royal Swedish Academy of Sciences in Stockholm in 1818. He served the Academy in this capacity for 30 years. In 1822, the American Academy of Arts and Sciences elected Berzelius as a Foreign Honorary Member. The Royal Society, London, awarded him the Copley Medal in 1836.

Berzelius' influence

Berzelius had far-reaching influence on the scientific scene in general, and chemistry in Sweden as well as in Europe.

In his own laboratory, he worked with a succession of young Swedish and foreign students who learnt his methods and thoughts first hand and spread them abroad when they left him. Most of them maintained close friendship with him. In his autobiographical notes, Berzelius lists 24 Swedes and 21 foreigners who worked in his laboratory. By the force of his personality, he exerted an influence which is still reflected in chemistry, more than a century after his death. Throughout his professional life, he kept track of the latest research being conducted in Europe and elsewhere and maintained contact with the leading researchers of his time. He became famous for his innovative experimental techniques and his laboratory attracted students from all over Europe. What made Berzelius unique was that he was committed to sharing his ideas, his methods of working and the results of his experimental research with others by publishing scientific articles and papers not only in Swedish but also in French, German and English. From 1821 to 1848, in his capacity as the perpetual secretary of the Royal Swedish Academy of Sciences, Berzelius published annual reports on the status of science in Swedish, German and French. These reports were highly popular and scientists across Europe looked forward to reading them. Berzelius was a builder of institutions. He enjoyed writing textbooks which were translated into other languages. They were translated into German by Wohler.

Berzelius was a meticulous experimentalist. He was a genius in planning his experiments to the minutest detail. His initial planning was so thorough that he seldom had to repeat the experiment as his preliminary preparation would have already suggested the result.

Berzelius is remembered as one of the most influential scientists of his era who played a crucial role in the development of modern chemistry. He strove throughout his professional career to bring about a holistic approach to chemistry and to unify the language of chemistry. He published more than 250 papers in his lifetime. Much like Lavoisier who initiated the change, he had the gift of summarizing the path breaking work of others and correlate the summarized data with his own research findings and arrive at a generalized theory to explain the experimental findings. His influence on modern chemistry is reflected even today in the 21st century.

Later years

Berzelius was somewhat conservative in his scientific outlook. As he grew older, newer developments occurred. The growth of organic chemistry changed the nature of chemistry and he found it difficult to accept some of them (for example, the radical theory of Liebig and Wohler, and the substitution principle of organic chemistry expounded by Dumas and Laurent) since they would destroy his dualistic theory.

As he advanced in age, Berzelius had indifferent health culminating in being paralysed in both his legs and was wheelchair-bound. Berzelius who operated at many levels and was engaged in many activities could no longer summon the strength to pursue his science. He died in Stockholm at his home on 7 August 1848 after a brief illness.

> Sweden acknowledges Berzelius' great contributions by calling him the Father of Swedish Chemistry and by annually celebrating August 20 as Berzelius Day.

(From www.stampboards.com)

A stamp issued by Sweden in 1979 honoring Berzelius

References

1. John Tyndall, *Sound: A Course of Eight Lectures Delivered at the Royal Institution of Great Britain* (1867), 26.
2. Johan Erik Jorpes and Barbara Steele, 'Jac.Berzelius: *His Life and Work"* University of California Press, 1970.
3. Bernard Jaffe; *Crucibles The story of chemistry from ancient alchemy to nuclear fission*, Chapter IX, p116 Dover publication, New York 1976.
4. Henry M. Leicester & Herbert S. Klickstein (Eds.), "*Essay on the Cause of Chemical Proportions and on Some Circumstances Relating to Them: Together with a Short and Easy Method of Expressing Them*". Annals of Philosophy 2 (1814) from *A Source Book in Chemistry*, 1400-1900, (Cambridge, MA: Harvard, 1952).
5. J. Black (trans.) "*An Attempt to Establish a Pure Scientific System of Mineralogy (1814),* 48.
6. *(Essai sur la théorie des proportions chemiques (1819)*, 98). "*Dictionary of Scientific Biography* (1981), *Vol. 2, 94*", & "Berzelius, Jöns Jacob." Complete Dictionary of Scientific Biography. 2008. *Encyclopedia.com.* 28 Jan. 2015 <http://www.encyclopedia.com>.

Jöns Jacob Berzelius, Source: http://www.chemistry.msu.edu/Portraits/images/Berzelius3c.jpg (From Wikipedia, The free encyclopedia, Jons Jacob Berzelius).

Chemicals on shelf in chemistry laboratory at Miami University (1911), Source: http://www.linkopingshistoria.se/1700-1850/1700-1780/

Image of a stamp depicting Berzelius at work in his laboratory, designed and engraved by Arne Wallhorn, and issued by Sweden on September 6, 1979. (Source: https://www.stampboards.com/)

5. MICHAEL FARADAY (1791–1867)

The greatest scientist of all time

"I think it will be conceded that Michael Faraday was the greatest experimental philosopher the world has ever seen; and I will add the opinion, that progress of future research will tend, not to dim or diminish, but to enhance and glorify the labours of this mighty investigator" (From Tyndall [1]).

"When an experimental result was obtained by Faraday it was instantly enlarged by his imagination. I am acquainted with no mind whose power and suddenness of expansion at the touch of new physical truth could be ranked with his" (From Tyndall [1]).

"The more we study the work of Faraday with the perspective of time, the more we are impressed by his unrivalled genius as an experimenter and a natural philosopher. When we consider the magnitude and extent of his discoveries and their influence on the progress of science and industry, there is no honour too great to pay to the memory of Michael Faraday - one of the greatest discoverers of all time" (From Thomas [2]).

(From Wikimedia Commons)

"Even if I could be Shakespeare, I think that I should still choose to be Faraday" (From Nugel, Sexton, Meckier [3]).

As remarked by Lord Porter, the life of Michael Faraday has a romantic 'rags to riches' quality and has remained a source of inspiration to all aspiring scientists as well as to laymen interested in stories of how science has been done by extraordinary human beings. Faraday's story is unique as he had to contend with every disadvantage that one can imagine - disadvantage of poverty, class (in a very class conscious England of the 18th century) and lack of formal education. That he could rise above such background and abject poverty, and become a colossus in the world of science illustrates the uniqueness of Faraday.

Early years

Faraday was born on 22 September 1791 in Newington Butts, a poor district on the outskirts of present-day London. Faraday's father was a poor blacksmith and his mother, daughter of a poor farmer. Providing formal schooling was never a priority as the family had to contend with lack of finances even for basic necessities. Faradays belonged to the Sandemanian sect – a breakaway sect of Christianity. Members of the sect did not hanker after materialistic comforts and worldly wealth and fame. Belief in the teachings of Christ perhaps helped the members of the sect to bear the hardships of life stoically. Faraday's outlook to life and science was influenced by these tenets. *"Faraday found no conflict between his religious belief and his activities as a scientist and philosopher..... A strong sense of the unity of God and nature pervades Faraday's life and work"* (From Baggot [4]).

Faraday attended a day school where he learnt the three Rs. Even this minimal opportunity for formal education was cut short by a series of unfortunate happenings at the school. The school believed in corporal punishment as an incentive to learning. Young Faraday had a serious speech problem and could not pronounce R's. This led to continuous corporal punishment at the hands of his teacher. After one particularly harsh punishment, Faraday collapsed to the floor and was unable to get up. His mother, summoned by his brother, was so outraged that she carried Michael out of the school once and for all. Much later, Faraday said of his education *"my education was of the most ordinary description, consisting of little more than the rudiments of reading, writing, and arithmetic at a common day-school. My hours out of school were passed at home and in the streets"* (From Hutchinson [5] & Jones [6]).

Apprenticeship with a bookbinder

By the time he was 13 years old, Faraday had to find a job to augment the family's meager resources. Faraday took up a job as an errand boy with a kind hearted book-binder, Riebau. Initially, Faraday's job was to deliver the newspapers on loan and bring them back to Riebau's shop. A year later Faraday became an apprentice book-binder. Encouraged by Riebau, Faraday read almost all the books pertaining to science that came for binding. Faraday recalled later *"Whilst an apprentice, I loved to read scientific books which were under my hands"* (From Jones [7]). Among the books that inspired and shaped his future scientific career were '*Encyclopaedia Britannica*' and '*Conversations on Chemistry*' by Jane Marcet. While the article on electricity in *Encyclopaedia Britannica* triggered Faraday's lifelong interest in electricity, Jane Marcet's book introduced him to the world of experimental science - chemistry. Faraday remarks about the profound influence of these two books: *"I was a very lively imaginative person, and could believe in the 'Arabian Nights' as easily as in the 'Encyclopaedia'. But facts were important to me, and saved me. I could trust a fact, and always cross-examine an assertion. So when I questioned Mrs. Marcet's book by such little experiments as I could find means to perform, and found it true to the facts as I could understand them, I felt that I had got hold of an anchor in chemical knowledge, and clung fast to it"* (From Tyndall [1]). However, Faraday did not let his lively imagination affect his intellectual pursuits. Tyndall remarks *"though the force of his imagination was enormous, he bridled it like a mighty rider, and never [let] his intellect [be] overthrown"* (From Tyndall [1]). Another book that influenced Faraday greatly and helped him to develop mental discipline *'The Improvement of the Mind'* by Isaac Watts.

Young Faraday conducted simple experiments to test chemical and physical concepts. He recalled that he *"made such simple experiments in chemistry as could be defrayed in their expense by a few pence per week, and also constructed an electrical machine.....*" (From Hutchinson [5]). He used discarded bottles and pieces of wood to build a crude but functioning electrostatic generator to conduct simple experiments in electricity. Later, he also built a voltaic pile stacking seven half pennies and seven disks of zinc (cut from a sheet of zinc) and six pieces of paper moistened with salt water. He used this device to decompose magnesium sulphate.

Faraday was constantly looking for avenues to educate himself. While he was serving his apprenticeship, he joined the *City Philosophical Society* established by a group of young men who were interested in self-improvement. The Society arranged evening lectures on topics related to natural philosophy (as science was called then). Faraday attended these lectures regularly and took down meticulous notes with drawings and bound them into books. Benjamin Abbott and Edward Magrath who became life-long friends were instrumental in developing his writing in English. While Abbott helped him to become skilful in written communication, Magrath helped with grammar, spelling and punctuation. He was fortunate to learn perspective drawing from a fellow lodger Masquerier, an artist. Riebau was proud of his young apprentice's efforts to educate himself and showed Faraday's bound books of lectures to his customers. Towards the end of his apprenticeship in 1812, one of Riebau's customers, William Dance, was so impressed by Faraday's interest in science that he gave Faraday four tickets to Sir Humphry Davy's last four lectures at The Royal Institution. This was the turning point in Faraday's career. Faraday was spellbound by Davy's oratory and even many years later, he could remember its impact. Faraday as his wont, took meticulous notes of Davy's lectures, embellished them with perspective drawings, bound them into a book and presented it to Davy. He was disappointed by Davy's lukewarm response.

Beginning of a scientific journey

When Faraday's apprenticeship with Mr. Riebau ended in 1812, he was desperate to change his profession from book-binding to science. His lack of formal education and an influential sponsor were however, serious obstacles. Faraday did not want to give up without trying his best. He wrote to the then President of the Royal Society (Sir Joseph Banks) to suggest a way for pursuing a scientific career, but did not receive a reply. He presented the bound book of the notes he had taken to Sir Humphry Davy, requesting him for a job. Davy sent Faraday away with a few kind words without any promise of a job. Davy advised Faraday to continue in his profession as a book-binder, saying *"Science is a harsh mistress, and in a pecuniary point of view (but) poorly rewarding those who devote themselves to her service"* (From Hammond [8]). Faraday was initially dejected, but refused to give up hope. In February 1813, came the proverbial lucky break. William Payne, a laboratory assistant at the Royal Institution was dismissed from his job for getting involved in a public fight in the Royal Institution's main lecture theatre. Davy was requested by the Royal Institution to find a replacement as soon as possible. After a brief interview, Davy recommended Faraday for the job.

Faraday was offered the job at a guinea per week with free accommodation of two rooms at the top of the Royal Institution building. He accepted thc job immediately even though the salary offered was lower than what he was earning as a book-binder. His main duty was to keep all the apparatus clean and ready so that Davy could conduct his experiments without the chore of washing glassware in the laboratory. In fact, the job description was 'bottle washer'.

Davy's recommendation letter presented to The Royal Institution on 18th March 1813

"*Sir Humphry Davy has the honour to inform the managers that he has found a person who is desirous to occupy the situation in the Institution lately filled by William Payne. His name is Michael Faraday. He is a youth of twenty-two of age. As far as Sir Humphry has been (able) to observe or ascertain, he appears well fitted for the situation. His habits seem good; his disposition active and cheerful, and his manner intelligent. He is willing to engage himself on the same terms as given to Mr. Payne at the time of quitting the institution*" (From Timmons [9]).

Thus began the journey that transformed Faraday from a book-binder's apprentice (with a minimal basic education and understanding of mathematics) into one of the greatest scientists of all time.

"*Sir H. Davy's greatest discovery was Michael Faraday*" (From Harvey [10]).

As Davy's assistant, Faraday helped Davy conduct experiments to separate elements in certain compounds by using electricity. These experiments gave Faraday an insight into the fundamental nature of electricity. By the time Faraday's job as Davy's assistant ended in 1820, Faraday had become an extraordinary experimentalist and had enough theoretical knowledge about electricity and other subjects to enable him to set up innovative experiments.

The Royal Institution was not merely his beloved laboratory but also his home from the time he entered it in 1813 till 1862 when he delivered his last Friday Evening lecture. Faraday felt that he owed all his successes in science to the generosity of the Royal Institution. He even turned down the offer of the Chair of Chemistry at the University of London in 1827 as he felt that he had much to contribute to making The Royal Institution stronger.

To Europe with Humphry Davy

In 1813, Davy decided to go on a long scientific tour to Europe. As his valet did not want to go, Davy asked Faraday to accompany him as his assistant and also to double up as his valet. Faraday, who had not travelled further than 12 miles from London, agreed to go on the European tour only when he was assured that he would get back his job at The Royal Institution on his return from the tour. The tour was exciting from the professional point of view as he helped Davy to conduct a number of experiments and meet famous European scientists. It was personally frustrating as Mrs. Davy, being an aristocrat, treated Faraday as a servant. When travelling by train, Faraday was made to sit outside the compartment and eat with the servants. He was so humiliated that he almost quit the tour but stayed on. The 18 months he spent travelling in Europe with

Humphry Davy provided Faraday the first opportunity to learn the intricacies of conducting experiments and broadened his scientific horizon. He learnt enough French and Italian to communicate with scientists in Europe. Faraday returned to the Royal Institution in April of 1815 a changed man and resumed his job. He was convinced that experiments were the only means to establish facts. He wrote on 19th April 1851 to John Tyndall *"Nothing is so good as an experiment which, whilst it sets an error right, gives us (as a reward for our humility in being reproved) an absolute advancement in knowledge"* (From James [11]).

"This year and a half (in Europe) may be considered as the time of Faraday's education; it was the period of his life that best corresponds with the collegiate course of other men who have attained high distinction in the world of thought. But his university was Europe; his professors were the master whom he served, and those illustrious men to whom the renown of Davy introduced the travelers. It made him personally known, also, to foreign savants at a time when there was little intercourse between Great Britain and the continent; and thus he was associated with the French Academy of Sciences while still young, his works found a welcome all over Europe, and some of the best representatives of foreign science became his most intimate friends" (From Gladstone [12]).

Faraday declared *"Without experiment I am nothing"* (From Jones [7]).

Faraday built simple devices to conduct complex experiments to advance understanding of difficult concepts. He believed that every theory or hypothesis or assertion had to be tested and verified by experiments. He said *"could trust a fact and always cross-question an assertion"* (From Tyndall [1]). It is for this reason he admired Priestley. He felt that all those who pursued science could learn from Priestley's freedom of thinking.

Faraday believed that Nature was the best teacher. He wrote to John Tyndall *"Nature is our kindest friend and best critic in experimental science if only we allow her intimations to fall unbiased on our minds..."* (From James [11]). *"One thing, however, is fortunate, which is, that whatever our opinions, they do not alter nor derange the laws of nature"* (From Jones [7]).

The year 1821 marked the beginning of a remarkable period in Faraday's personal and professional life. Professionally, it marked Faraday's focus shift from chemistry to physics of electricity and magnetism. On the personal front, he married Sarah Barnard a fellow Sandemanian who was a life-long source of strength to him.

Chemist, physicist or electrical engineer?

The genius of Faraday and his unparalleled contributions to all these three fields were so enormous that each of them claimed him as their own.

Faraday the chemist: *"There is in the chemist, a form of thought by which all ideas become visible in the mind as strains of an imagined piece of music. This form of thought is developed in Faraday in the highest degree"* (From Brock [13]).

Faraday possessed 'chemical intuition' in abundance which chemists claim as their special quality that helps them to leapfrog from a mere idea to conclusions by experimental verification. He started his scientific career as a chemist under Davy, one of the greatest experimental chemists. His phenomenal contributions to chemistry include making steel alloy (1822), liquefaction of chlorine gas (1823), discovery of benzene which he named bicarburet of hydrogen (1825), synthesizing compounds of carbon and chlorine for the first time, introducing the system of oxidation numbers, determination of chlorine hydrate's composition (which Humphry Davy had discovered in 1810). Faraday became one of the best chemical analysts of his time.

Faraday formulated the laws of electrolysis. It is a wonder how he could formulate the laws when the nature of an atom was not even well understood at that time. Electrons had not yet been discovered.

Helped by his friend Whewell, Faraday popularized terms like anode and cathode derived from Greek- ana meaning 'up' and Kata meaning 'down' and hodos meaning road; hence, anode was the up road and cathode the down road; ion meaning 'wanderer' in Greek hence 'anion' and 'cation'. Other important terms coined were 'electrolyte' and 'electrode'. These terms are used even today.

"I require a term to express those bodies which can pass to the electrodes....... I propose to distinguish these bodies by calling those anions which go to the anode of the decomposing body; and those passing to the cathode, cations; and when I have occasion to speak of these together, I shall call them ions" (From Faraday [14]).

Laws of Electrolysis: Faraday wrote to Eilhard Mitscherlich on 24 Jan 1838, *"I have been so electrically occupied of late that I feel as if hungry for a little chemistry: but then the conviction crosses my mind that these things hang together under one law & that the more haste we make onwards each in his own path the sooner we shall arrive, and meet each other, at that state of knowledge of natural causes from which all varieties of effects may be understood & enjoyed"* (From James [15]).

Faraday conducted a series of experiments to study the relation between electricity and chemistry, particularly electrochemical reactions. He found that the quantity of elements separated from a dissolved or molten salt when an electric current was passed through it, was directly proportional to the quantity of current passing through the circuit. On the basis of this observation, he formulated the famous Laws of Electrolysis in 1834.

Herman Von Helmholtz in his Faraday lecture in 1881 said *"The most startling result of Faraday's Law is perhaps this: If we accept the hypothesis that the elementary substances are composed of atoms, we cannot avoid concluding that electricity also, positive as well as negative, is divided into definite elementary portions, which behave like atoms of electricity"* (From Helmholtz [16] & Datta [17]). Electrochemical industry is governed by these laws even today.

Faraday the first nanoscientist: Faraday in his Bakerian Lecture in 1857 titled 'Experimental Relations of Gold (and other Metals) to Light' said that *"(when) the gold is reduced in exceedingly fine particles which becoming diffused, produce a beautiful fluid ....the various preparations of gold whether ruby, green, violet or blue ......consist of that substance in a metallic divided state"* (From Faraday [18]) and further observed that the optical properties of the substance in metallic divided state were different from those of bulk gold. The relation between particle size and chemical properties or the effect of quantum size on properties of nanoparticles was observed for the first time!

Faraday had a holistic view of chemistry. He conducted experiments in different branches of the subject and had scant respect for those who compartmentalized the subject. He wrote to William Grove *"I think chemistry is being frittered away by the hairsplitting of the organic chemists; we have new compounds discovered, which scarcely differ from the known ones and when discovered are valueless—very illustrations perhaps of their refinements in analysis, but very little aiding the progress of true science"* (From Kahlbaum & Darbishire [19]). Faraday worked on a number of problems in different areas simultaneously.

Faraday's major contributions to chemistry

(From Thomas [2] & Rao [20])

1816 Miner's safety lamp (with Humphry Davy),

1818-24 Preparation and properties of alloy steel,

1812-30 Analytical chemistry work- on clays, native lime, water, gun powder, rust, various gases, liquids and solids, electrochemistry,

1820-26 Organic chemistry-discovery of benzene, isobutylene, tetrachloroethylene, hexachlorobenzene, isomers of naphthalene-sulfonic acids, photochemical reactions,

1825-31 Production of optical glass,

1823, 1845 Liquefaction of gases, existence of critical temperature and continuity of state,

1833-36 Electrochemistry- laws of electrolysis etc,

1834 - Heterogeneous catalysis, surface reactions, adsorption, wettability of solids. Faraday effect, Faraday cage, Faraday constant, Faraday cup, Faraday's laws of electrolysis, Faraday paradox, Faraday rotator, Faraday-efficiency effect, Faraday wave, Faraday wheel and Lines of force.

Physicist or chemical physicist? Faraday's seminal contributions to electricity and magnetism are unparalleled. He created new fields, experimentally proved the relationship between electricity and magnetism and yet he did not want to be called a physicist! Faraday felt that *"The new term Physicist is both to my mouth and ears so awkward that I think I shall never use it. The equivalent of three separate sounds of i in one word is too much"* (From Ross [21]). He preferred to be called a natural philosopher.

After Hans Christian Oersted made the exciting discovery of the relationship between electricity and magnetism in 1820, Faraday then had started to conduct research in this new and exciting field at the insistence of Richard Philips, one of the editors of *Philosophical Magazine.* Faraday's mentor Humphry Davy and his friend William Hyde had failed in their attempts to build an electric motor. Faraday discussed the problem with them and decided to repeat Oersted's experiment to understand the phenomenon and conducted a series of experiments in the basement of the Royal Institution. And the rest is history.

Electromagnetic rotation

Faraday's experiment was remarkable for its simplicity. He fixed a magnet upright to the bottom of a deep basin in such a way that only one of the poles of the magnet was above the surface. The basin was filled with mercury. A wire which was free to move was suspended above the bowl and was dipped into the mercury. When a current was passed through the wire, the wire rotated around

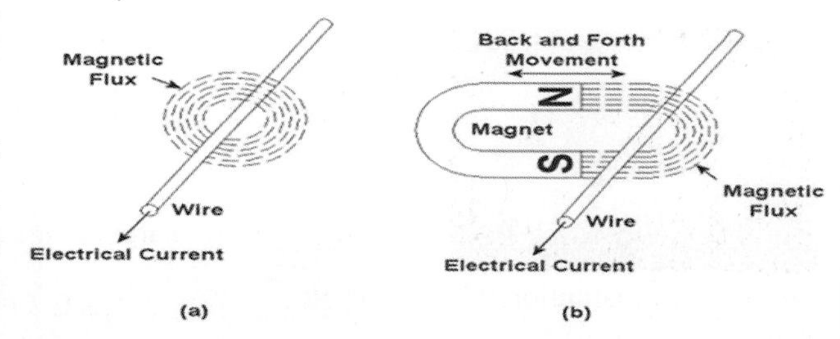

the fixed magnet continuously as long as the current was passing through it. He then fixed the wire and left the magnet free to move and passed current through the wire. The magnet rotated around the wire as long as current was flowing through the wire. These simple experiments resulted in his discovery of electro-magnetic rotation and demonstrated for the first time that in both cases, electrical energy was being converted to mechanical energy. Faraday argued correctly that the circular rotation of both the magnet and the wire was due to the magnetic field pushing them. He called this motion 'electromagnetic rotation' - the basis of an electric motor. With this simple and innovative experiment, Faraday discovered the fundamental concepts of magnetic lines of force and magnetic field and laid the foundation for today's electromagnetic technology.

Faraday was so excited by the success of his device that he published the result without acknowledging Davy and William Hyde Wollaston. Davy accused Faraday of cheating and this affected Faraday so deeply that he decided to quit conducting further research in this field. He again started working in this field only in 1831, two years after Davy's death in 1829.

Electromagnetic induction: On 29 August 1831, Faraday discovered electro-magnetic induction - the principle behind the electric transformer and generator. Following this discovery, Faraday wrote to Richard Phillips on 23 Sep 1831 *"I am busy just now again on electro-magnetism and think I have got hold of a good thing but can't say; it may be a weed instead of a fish that after all my labour I may at last pull up"* (From Jones [7]).

From Royal Institution (C) Paul Wilkinson 2010

Michael Faraday's Induction Ring

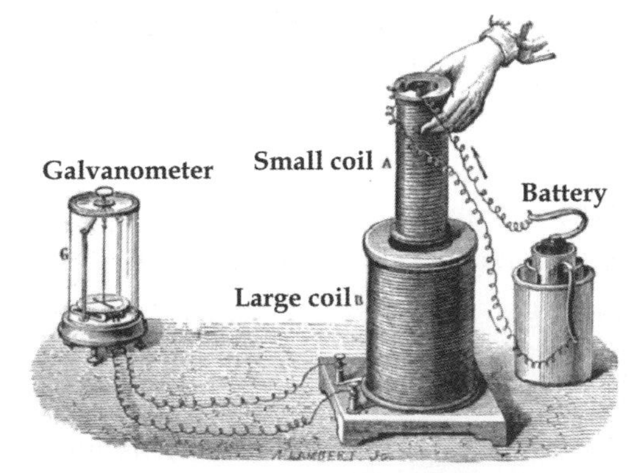

(From 'Michael Faraday', *Wikipedia*)

Faraday's 1831 experiments demonstrating induction

In 1821, Faraday had shown that it was possible to convert electrical energy into motion and ten years later in 1831, he demonstrated that the opposite (conversion of mechanical energy to electrical energy) was also possible.

Faraday's lack of knowledge of advanced mathematics was a blessing in disguise. His powerful visual imagination enabled him to mentally see the lines of force and present his discovery of the force of magnetic fields not with complex mathematical equations but with elegant drawings. It laid the foundation for one of the four famous equations of Maxwell and the exciting field theory. Maxwell considered Faraday as the founder of electro-magnetism.

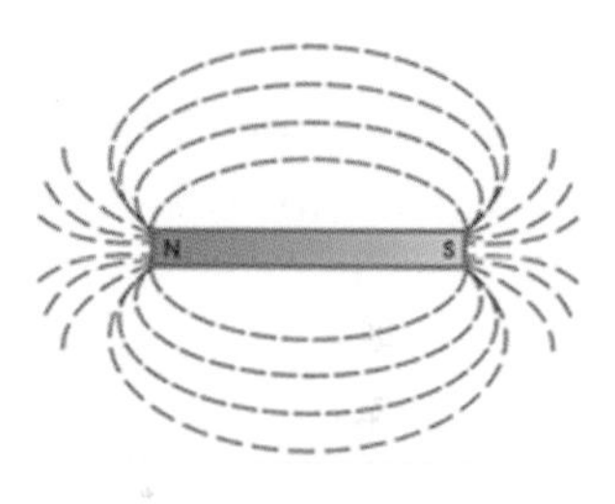

He remarked *"Magnetic lines of force convey a far better and purer idea than the phrase magnetic current or magnetic flood; it avoids the assumption of a current or of two currents and also of fluids or a fluid, yet conveys a full and useful pictorial idea to the mind"* (From Martin [22]).

Maxwell said "*Faraday is, and must always remain the father of that enlarged science of electro-magnetism*" (From Maxwell [23]).

Incredible 1830: Timeline of Faraday's mindboggling discoveries in physics and electricity

April 2 - the power of Electricity to confer phosphorescence on certain metals

April 29 - On Mr. Trevelyan's recent experiments on the production of sound during the conduction of heat

May 12 - on a peculiar class of acoustical figures and on certain forms assumed by groups of particles upon vibrating elastic surfaces

June 18 - on the arrangements assumed by particles on the surfaces of vibrating elastic bodies

Nov 24 - Experimental researches in Electricity

1) On the induction of Electric currents
2) On the evolution of electricity from Magnetism
3) On a new electrical condition of matter
4) On Arago's magnetic phenomena

Faraday's impact on electricity and related fields

On 24 November 1831 and 12 January 1832, Faraday presented two papers to the Royal Society in which he describes his experiments in detail. They form the first and second parts of his '*Experimental Researches into Electricity*' where Faraday formulates the law which governs the evolution of electricity by magneto-electric induction. This discovery was instrumental in changing the nineteenth century view of electricity as a mere scientific curiosity into a storehouse of technology that irreversibly changed the world.

"Electricity is often called wonderful, beautiful; but it is so only in common with the other forces of nature. The beauty of electricity or of any other force is not that the power is mysterious, and unexpected, touching every sense at unawares in turn, but that it is under law, and that the taught intellect can even govern it largely. The human mind is placed above, and not beneath it" (From Jones [7]).

In the 1830s, Faraday conducted a series of experiments to understand the fundamental nature of electricity. Through these experiments, he proved that no matter how electricity was produced – induced by a magnetic field, produced by a voltaic pile or any other type of chemical battery, by static electricity or by animals like the electric eel – only a single electricity exists. He established that electricity was not an imponderable fluid as it was believed at that time but a force which flowed from particle to particle in a conductor.

Static electricity and the Faraday cage

Towards the latter half of the 1830s, Faraday took up the study of static electricity. He observed that when a charge was applied to a conducting container on the outside, there was no electrical charge inside the closed container. This observation led him to conclude correctly that this was due to the charge being redistributed inside the container to cancel the charge applied on the outer surface of the container thus creating a shielding effect. This effect offers protection from atmospheric phenomena like electrostatic discharges and lightning strikes. This shielding effect is what protects the occupants of a car from the lethal effect of a lightning strike.

Ice pail experiment

Faraday's device to test the nature of static electricity consisted of a hallow metallic pail (generally used to keep ice), a wooden stool for grounding the pail to remove the excess charge, a charged metal ball hung from a non conducting silk thread, a gold leaf electroscope attached by wire to the outer surface of the pail. Faraday showed that when the charged ball was lowered carefully without touching the sides, the charge was registered by the electroscope but when the ball touched the side of the container, the electroscope did not register any charge. This demonstrated the nature of static electricity. As Faraday used an ice pail for conducting the experiment, the experiment came to be known as Ice pail experiment. This simple experiment can be used to demonstrate the principles of electrostatics.

In 1839, Faraday's health broke down as a result of eight years of intense involvement with research and the strain of designing and building devices for conducting experiments. His creative work again started only in 1845.

Paramagnetism, Diamagnetism and Faraday Rotation: *"I have at last succeeded in illuminating a magnetic curve or line of force and in magnetising a ray of light"* (From Day [24]).

According to John Tyndall, Faraday's *"third great discovery is the magnetization of light, which I should liken to the Weisshorn among mountains- high, beautiful and alone"* (From Tyndall [1]).

When Faraday resumed active research in 1845, *"he concentrated on finding proof for his belief in the existence of the all encompassing universal force"* (From Baggot [4]). At the suggestion of young William Thomson (who later became Lord Kelvin), Faraday conducted his famous experiment in which he passed a plane-polarized beam of light through a special optical glass with very high refractive index which he had developed in the 1820s. When he switched on an electromagnet whose lines of force were parallel to the light ray, the plane of polarization rotated. Faraday further observed that even when the direction of the ray of light was changed, the rotation remained in the same direction. Faraday deduced correctly that the direction of the rotation of the plane of polarization of the beam of light was due solely to the polarity of the magnetic lines of force and the optical glass only helped to detect this change in direction. The effect of magnetism on light was observed for the first time. This discovery convinced him that all matter would respond in some way to a magnetic field. He experimented with a variety of materials such as nickel, cobalt, and oxygen as well as some aqueous solutions and gases by placing them between the poles of an electromagnet. He found that while some materials such as iron aligned themselves along the lines of force and were therefore attracted by the more intense parts of the electromagnetic field, some like bismuth aligned themselves across the magnetic field and therefore moved towards the less intense magnetic field. Faraday called the first group of materials '*paramagnetics*' and the second group '*diamagnetics*'. Further research helped Faraday to conclude that paramagnetic materials were better conductors of magnetic lines of force of the surrounding medium than the diamagnetic materials.

Faraday's experiment to prove oxygen was paramagnetic

On learning about the discovery in 1847 by an Italian scientist that flames responded to a magnetic field, Faraday wondered if gases would also be affected by a magnet (or a magnetic field). He conducted a simple experiment to test the behaviour of air in a magnetic field. He made soap bubbles containing oxygen and nitrogen and passed them through a magnet (magnetic field). While the soap bubbles containing oxygen moved towards the magnet, nitrogen containing soap bubbles went through without any deviation, proving that oxygen was paramagnetic (From Tyndall [1]).

This discovery of the effect of magnetic field on light and other materials provided the proof for his long held conviction regarding the unity of forces of Nature. Faraday wrote to his friend Christian Schöenbein on 13 November 1845 *"I happen to have discovered a direct relation between magnetism and light, also electricity and light, and the field it opens is so large and I think rich"* (From Kahlbaum & Darbishire [19]).

> This astonishing discovery was announced to the world on November 8th, 1845, in the magazine *Athenaeum* thus: *"Mr. Faraday, on Monday (November 3rd), announced at a meeting of the council of the Royal Institution a very remarkable discovery, which appears to connect the imponderable agencies yet closer together, if it does not indeed prove that light, heat and electricity are merely modifications of one great universal principle"* (From Jones [7]). This effect is known now as *Faraday Effect* and the rotation of light as *'Faraday Rotation'*.

Unification

In 1846, a curious and unexpected incident led Faraday to speculate in public about his view of the unified forces of Nature. The lecturer scheduled to give the Friday evening public lecture at the Royal Institution got stage fright at the last minute and ran away from the scene. As the hall was packed and with no scheduled lecturer in sight, Faraday had no option but to step in. In his lecture titled 'Thoughts on Ray Vibrations', he spoke eloquently about atoms and their lines of force. He further suggested that the light waves were propagated via (the medium of) lines of magnetic and electrical forces associated with the atoms. This speculation laid the foundation for Maxwell's theory of electromagnetic fields. It never ceases to astonish the modern day scientists how Faraday with a mere three years of formal schooling and with no mathematical background could arrive at this fundamental conclusion when the concept of atoms and lines of force and field were unknown! Almost towards the end of an illustrious career, Faraday proposed a radical view of the relationship between space and force- that electromagnetic forces extended into the empty space around the conductor. According to this, space was not just where bodies

and forces were located but was actually a medium with a capacity to support magnetic and electric forces. This can be regarded as the birth of field theory. Brian Bowers in his book 'Michael Faraday and Electricity *(Pioneers of science and discovery)'* writes *"it seems likely that his religious belief in a single Creator encouraged his scientific belief in the "unity of forces", the idea that magnetism, electricity and the other forces have a common origin"* (From Bowers [25]).
Einstein had photographs of Faraday and Maxwell in his office.

Honours

"I have always felt that there is something degrading in offering rewards for intellectual exertion, and that societies or academies, or even kings and emperors should mingle in the matter does not remove the degradation" (From James [26]).

Faraday's views on honours and awards, especially those conferred by his own country were in direct contrast to those of his mentor Davy. He refused to accept the presidency of the Royal Society not once but twice. On being offered the Presidency of the Royal Society, he wrote to his friend John Tyndall *"I must remain plain Michael Faraday to the last; and let me now tell you, that if accepted the honour which the Royal Society desires to confer upon me, I would not answer for the integrity of my intellect for a single year"* (From Tyndall [1]). For the same reason when Queen Victoria offered him a knighthood, he politely refused it and also turned down the honour of being buried in Westminster Abbey.

Faraday strongly believed that pursuit of science was a reward in itself. He felt that science (education) *"teaches us to deduce principles carefully, to hold them firmly, or to suspend the judgment, to discover and obey law, and by it to be bold in applying to the greatest what we know of the smallest. It teaches us first by tutors and books, to learn that which is already known to others, and then by the light and methods which belong to science to learn for ourselves and for others; so making a fruitful return to man in the future for that which we have obtained from the men of the past"* (From Jones [7]).

Faraday was honoured for his outstanding contributions to chemistry, physics and electricity (which he perhaps unwillingly accepted).

Faraday received many awards and honours. Some of them are: Fellow of the Royal Society (1825), Honoris Causa by the University of Oxford (1832), Fullerian Professorship for life (1833), Copley Medal (1832), Royal Medal in Chemistry (1835), Copley Medal (1838) of the Royal Society, Foreign member of the Royal Swedish Academy of Sciences (1838), French Academy of Sciences (One of the eight foreign members at that time, 1844), Royal Medal of the Royal Society in physics and Rumford Medal of the Royal Society (physics) (1846).
He is one of the few scientists to receive both The Royal Medal and the Copley Medal twice.

Faraday was totally disinterested in acquiring wealth through scientific discoveries. His research was totally focused on understanding the true nature of scientific phenomena. Almost every one of his discoveries contributed to path changing technologies and yet at no time was he personally interested in their utilitarian value. A typical example of this was his research on electricity. It is often quoted that when the Chancellor of the Exchequer, William Gladstone, asked Faraday what was the practical use of his discovery of electricity, Faraday apparently replied that he was not sure of it at present *"but Sir there is every probability that you will soon be able to tax it!"* (From Lecky [27]). When the Prime Minister asked of a new discovery made by Faraday, 'What good is it?' Faraday apparently replied, *"What good is a new-born baby?"* (From Cohen [28]).

Work ethics

Faraday's advice to William Crookes, a student, was *"The secret (of success) is comprised in three words — Work, finish, publish"* (From Gladstone [12]).

Faraday worked every day, maintained the same rigorous schedule and maintained meticulous notes of the work done at his laboratory. He worked every day. In 1833, his notebook had entries for 24th December and 26th December. There was no entry on the 25th for the obvious reason (Christmas day).

Faraday felt the five essential skills required for success were "concentration, discrimination, organization, innovation and communication". He had these skills and the required mental discipline in ample measure. Faraday wrote to C. Ransteed *"I have never had any student or pupil under me to aid me with assistance; but have always prepared and made my experiments with my own hands, working & thinking at the same time. I do not think I could work in company, or think aloud, or explain my thoughts at the time. Sometimes I and my assistant have been in the Laboratory for hours & days together, he preparing some lecture apparatus or cleaning up, & scarcely a word has passed between us; - all this being a consequence of the solitary & isolated system of investigation; in contradistinction to that pursued by a Professor with his aids & pupils as in your Universities"* (From Williams [29]).

Evening Discourses and Christmas lectures

Faraday was committed to popularising scientific knowledge among the general public perhaps as he himself did not have the benefit of education in his youth. To achieve this, he established in 1826 Friday Evening Discourses and later in the same year, the Christmas Lectures at the Royal Institution. Through these lectures, he wanted to educate and inspire the young by first capturing their interest by entertaining them with simple

(From "Michael Faraday", Wikipedia)

experiments underlying the fundamental laws of physics and chemistry. Faraday passionately believed that *"lecturing is capable of improving not only those who are lectured, but also the lecturer*" (From Jenkins [30]). *A lecturer should exert his utmost effort to gain completely the mind and attention of his audience, and irresistibly make them join in his ideas to the end of the subject........... A flame should be lighted at the commencement and kept alive with unremitting splendour to the end"* (From Jones [7]). He gave nineteen lectures in all and of these, only the lecture series *'A Chemical History of a Candle'* (1848) and '*The Various Forces of Matter and their Relations to Each Other*' (1859) have survived in entirety. The themes of these lectures are centered on Faraday's respect for laws of Nature and his conviction about the unity of the fundamental forces of Nature. He felt that the study of physical sciences was best suited and '*more capable of giving him (man) an insight into the actions of those laws, a knowledge of which gives interest to the most trifling phenomenon of nature, and makes the observing student find'* - "*Tongues in trees, books in the running brooks, Sermons in stones, and good in everything*?" (From Crookes [31]).

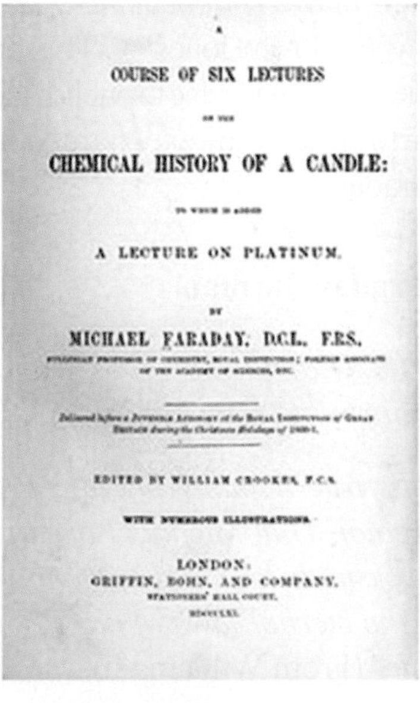

A Chemical History of Candle: *"I purpose........to bring before you, in the course of these lectures, the Chemical History of a Candle. There is not a law under which any part of this universe is governed which does not come into play, and is touched upon in these phenomena. There is no better, there is no more open door by which you can enter the study of natural philosophy, than by considering the physical phenomena of a candle"- Michael Faraday* (From Faraday [32]). He gave six lectures in all covering every aspect of the chemistry and physics of flames.

In 1861, these lectures were published as a book of the same title. In his concluding remarks for the sixth and final lecture at the Royal Institution (Christmas 1860-61) Faraday exhorted the children

"I ... express a wish that you may, in your generation, be fit to compare to a candle; that you may, like it, shine as lights to those about you; that, in all your actions, you may justify the beauty of the taper by making your deeds honourable and effectual in the discharge of your duty to your fellow-men" (From Day [24] & Carey [33]).

Public service

Faraday was perhaps one of the earliest environmentalists. During a boat trip he took down the river Thames in 1855, Faraday noticed the extreme pollution level of the river and wrote the famous letter to The Times newspaper. "....... *surely the river which flows for so many miles through London ought not to be allowed to become a fermenting sewer"* (From Jones [7]). This prompted The Punch magazine to publish the famous cartoon of Father Thames greeting Faraday.

(From "Michael Faraday", Wikipedia)

Faraday the man

"His [Faraday's] soul was above all littleness and proof to all egotism" (From Tyndall [1]).

"His true humility lay in a profound consciousness of his debt to his Creator. That Michael Faraday, poor uneducated son of a journeyman blacksmith and a country maid was permitted to glimpse the beauty of the eternal laws of nature was a never-ending source of wonder to him" (From Williams [34]).

Faraday's lively imagination and ability to create mental picture perhaps was responsible for Faraday's ability to build, as Maxwell put it, a mental picture of lines of force, filling space, shaping themselves into lovely arrays. *"His second great characteristic was his imagination. It rose sometimes to divination, or scientific second sight, and led him to anticipate results that he or others afterwards proved to be true"* (From Jones [7]).

Faraday and the Nobel Prize: By all accounts and according to many outstanding scientists including Lord Rutherford, had there been Nobel Prizes during his lifetime, Faraday would have been awarded at least six to seven prizes – an unbeatable all time record. But knowing his views on scientific awards, in all probability he would have politely refused the honour.

Last years

"There was a philosopher less on earth, and a saint more in heaven" Gladstone's tribute to Faraday (From Gladstone [12]).

In 1861 while giving a lecture at the Royal Institution, Faraday suffered a memory lapse and lost his train of thought. After this incident, he decided to resign from his job as he feared this would increase and become a source of embarrassment both to himself and to his beloved institution. He retired to the cottage at Hampstead - Queen Victoria's gift in recognition of his services to science. His dementia progressively worsened until he passed away peacefully in the company of his devoted wife on August 25, 1867. In accordance with his wish, he was buried in Highgate Cemetery in a plot belonging to the Sandemanian Society. He remained a kind, humble and socially self effacing person till the end unaffected by all the admiration of fellow scientists and general public.

The following excerpts from the obituary tribute in *The Times* best sums up the extraordinary scientist and human being - one and only Michael Faraday.

"The world of science lost on Sunday one of its most assiduous and enthusiastic members. The life of Michael Faraday had been spent from early manhood in the single pursuit of scientific discovery, and

through his years extended to 73, he preserved to the end the freshness and vivacity of youth in the exposition of his favourite subjects, coupled with a measure of simplicity which youth never attains............. as a man of science he was gifted with the rarest of felicity of experimenting, so that the illustrations of ... subjects seemed to answer with magical ease to his call. ...... He was one of those men who have become distinguished in spite of every disadvantage of origin and of early education, and if the contrast between the circumstances of his birth and of his later worldly distinction be not so dazzling as is sometimes seen in other walks of life, it is also true that his career was free from the vulgar ambition and uneasy strife after place and power which not commonly detract from the glory of the highest honours. No man was ever more entirely unselfish, or more entirely beloved. Modest, truthful, candid, he had the true spirit of a philosopher and of a Christian, for it may be said of him, in the words of the father of English poetry,—

"*Gladly would he learn, and gladly teach.*"

The cause of science would meet with fewer enemies, its discoveries would command a more ready assent, were all its votaries imbued with the humility of Michael Faraday" (From The Times [35]).

References

1. John Tyndall, "*Faraday as a Discoverer*" (4th Edition), Longmans, Green, London, 1868.

2. J. M. Thomas, *Michael Faraday and The Royal Institution: The Genius of Man and Place*, 1991, p129.

3. Bernfried Nugel, James Sexton, Jerome Meckier (Eds.), Aldous Huxley Annual: Volume 9 (2009), (2011).

4. Jim Baggot, "The myth of Michael Faraday....", *New Scientist No.1787, (21 September 1991)*).

5. Ian H. Hutchinson, *"The Genius and Faith of Faraday and Maxwell,"* The New Atlantis, Number 41, Winter 2014, p. 81-99.

6. Bence Jones, Michael Faraday, *The Life and Letters of Faraday*, 2010.

7. Henry Bence Jones, *The Life and Letters of Faraday Vol. II, p. 483, Cambridge University Press, 1870.*

8. D. B. Hammond, *Stories of Scientific Discovery,* Cambridge University Press, 1923.

9. Todd Timmons: *Makers of Western Science: The Works and Words of 24 Visionaries from Copernicus to Watson and Crick*, p 128, (2012).

10. 'Michael Faraday', in Paul Harvey (ed.), *The Oxford Companion to English Literature,* Oxford, 1932.

11. *Frank A. J. L. James (Ed.),* letter 2411, *The correspondence of Michael Faraday, Volume 4.* The Institution of Engineering and Technology, London, *1999.*

12. J. H. Gladstone, *Michael Faraday,* New York Harper & Brothers, 1872.

13. *Autobiography*, Justus von Liebig, 257-358, Quoted in William H. Brock, *Justus Von Liebig,* Cambridge University Press; 2002.

14. Faraday, M., *Experimental Researches in Electricity. Seventh Series.* Philosophical Transactions of the Royal Society of London (1776-1886), 1834, 124, 79.

15. (Letter to Eilhard Mitscherlich, 24 Jan 1838), Frank A. J. L. James (ed.), *The Correspondence of Michael Faraday* Vol. 2, 488, The Institution of Electrical Engineers, London 1993.

16. *Helmholtz's Faraday Lecture April 5, 1881* before the Fellows of the Chemical Society, London.

17. N. C. Datta, *'The Story of Chemistry',* Universities Press, 2005, pg 206.

18. *The Philosophical Transactions of The Royal Society of London,* Michael Faraday, vol.147 *(1857), pp 145-187,* The Royal Society.

19. *Georg W. A. Kahlbaum and Francis V. Darbishire (Eds.),* (5 Jan 1845, Letter to William Grove) *The Letters of Faraday and Schoenbein, 1836-1862, London,* 1899.

20. CNR Rao, 'Understanding Chemistry', Universities Press (1999).

21. Sydney Ross, *Nineteenth-Century Attitudes: Men of Science, Chemists and Chemistry,* (1991), p10 Kluwer Academic Publishers.

22. Thomas Martin (ed.), Diary Entry for 10 Sep 1854, *Faraday's Diary: Being the Various Philosophical Notes of Experimental Investigation* (1935), Vol. 6, 315.

23. James Clerk Maxwell, *Scientific Worthies I.* - Faraday, Nature 8 (1873), p 400 and 398.

24. Peter Day, *(Entry in Faraday's notebook),* (1999). *The Philosopher's Tree: A Selection of Michael Faraday's Writings.* CRC Press, p125.

25. Brian Bowers, 'Michael Faraday and Electricity *(Pioneers of science and discovery)'* Priory Press, 1974, Edition 2.

26. Michael Faraday, Frank AJL James (Ed.), *The Correspondence of Michael Faraday, Volume 3: 1841-1848,* Letter 1466. The Institution of Electrical Engineers. 1996.

27. *As quoted in W.E.H. Lecky, Democracy and Liberty (1899).*

28. I. Bernard Cohen, *Benjamin Franklin's Science,* 1996.

29. L. Pearce Williams (Ed.), Faraday's letter of 16 Dec 1857, '*The Selected Correspondence of Michael Faraday'* (1971), Vol. 2.

30. Alice Jenkins (Ed.), *Michael Faraday's Mental Exercises: An Artisan Essay-circle in Regency London,* Liverpool University Press, 2008.

31. Michael Faraday, William Crookes (Ed.) *A course of six lectures on the Various Forces of Matter and their Relations to Each Other,* Harper & Brothers, Publishers, New York, 1860.

32. Michael Faraday, *The Chemical History of a Candle,* 1861.

33. John Carey (Ed.), *Eyewitness to Science*, 1995.

34. Williams, L.P. *Michael Faraday*, Simon and Schuster, 1971.

35. Excerpts from the Obituary published in The Times, Wednesday, Aug 28, 1867; Issue 25901, pg.7; col C- The Late Professor Faraday.

Photograph of Michael Faraday (1861), Source: Opposite p. 290 of R. A. Millikan and H. G. Gale's "Practical Physics" (1922) (Wikimedia Commons)

Faraday's 1831 experiments demonstrating induction. 1892, From Arthur William Poyser (1892) *Magnetism and electricity: A manual for students in advanced classes*, Longmans, Green, & Co., New York, p. 285, fig. 248 on Google Books. The drawing is signed *Lambert, J. (From Wikipedia, the free encyclopedia).*

Michael faraday's Induction Ring, Royal Institution (C) Paul Wilkinson 2010.

Detail of a lithograph of Michael Faraday delivering a Christmas lecture at the Royal Institution, 1856 by Alexander Blaikley. (Source: *Notes and Records of the Royal Society of London*, 2002, volume 56, page 370) *(From Wikipedia, the free encyclopedia).*

Michael Faraday meets Father Thames, from *Punch* (21 July 1855)
John Leech Cartoons from *Punch* magazine, retrieved 3 May 2013 (Wikimedia Commons).

6. FRIEDRICH WÖHLER (1800–1882)

The one who made the first organic compound

From "Friedrich Wöhler", Wikipedia

"For two or three of his researches, he deserves the highest honor a scientific man can obtain, but the sum of his work is absolutely overwhelming. Had he never lived, the aspect of chemistry would be very different from that it is now" (From Scientific American Supplement [1], 1882).

Early years

Friedrich Wöhler was born to Anton August Wöhler and Anna Katharina Schröder, daughter of a professor of philosophy on 31 July 1800 in Eschersheim (now a district of Frankfurt am Main). His father, following his family tradition, started as an equerry to the elected representatives of the district of Hesse. He quit it to take up farming in Frankfurt and went on to become a prominent citizen of that city while his mother devoted herself to the family. His father was Friedrich Wöhler's first teacher instructing him in writing and drawing.

Young Wöhler started his education at a public school when he was about seven years old. His father being himself well educated arranged for instruction in Latin, French as well as in music to prepare his son for studying in the Gymnasium at Frankfurt. He joined the Gymnasium in 1814. Young Wöhler was influenced by one Dr. Buch, who was interested in the pursuit of science as a hobby. On special occasions, he allowed Wöhler to conduct experiments in his kitchen cum laboratory. Wöhler later credited Dr. Buch for encouraging him to develop interest in chemistry and passion for conducting chemical experiments. After graduating from the Gymnasium in 1820, he joined the university at Marburg where he chose to study medicine. He moved to the University of Heidelberg a year later where he came under the influence of Gmelin, professor of chemistry and medicine. Wöhler was encouraged by Gmelin and his colleague Tiedemann to pursue physiological studies which proved to be an asset in his later career. His thesis on 'Migration of Matter into Urine,' was adjudged the best thesis in the competition held by the faculty of medicine. This study proved helpful in later research. Wöhler graduated from the University of Heidelberg with the Doctor of Medicine and Surgery, *"nee non artis obstetriciæ"* degree in 1823. Wöhler was persuaded by Gmelin to give up an uncertain career in medicine and opt for a career in chemistry. Under Gmelin's guidance, Wöhler started his research in sulphocyanic acid and cyanic acid. The results of his work were published in Annalen der Physik or Gilbert's Annalen (one of the oldest journals published since 1799). Gmelin advised Wöhler to work under Berzelius to further hone his laboratory skills and strongly recommended Wöhler to his close friend Berzelius. Berzelius who held Gmelin in high regard wrote back to Wöhler accepting him as a student. "*Any one who has studied chemistry under the direction of Leopold Gmelin has very little to learn from me. Nevertheless, I cannot forego the pleasure of making your personal acquaintance, and will therefore cheerfully accept you as the companion of my labors. You can come whenever it is agreeable to you*" (From Joy [2]).

Berzelius was a hard taskmaster. The rigorous and meticulous training in analysis of minerals that he received under Berzelius for a year helped Wöhler to sharpen his experimental skills to a great extent. Also, Wöhler formed a long lasting friendship with Berzelius which lasted till Berzelius' death in 1848. Over the years, Wöhler translated many of Berzelius's works, including all the six parts of his famous textbook of chemistry into German.

Wöhler returned to Germany in 1825 and started his academic career at an industrial school in Berlin. In 1831, he moved to Kassel as a professor. In 1836, he accepted the position of professor of chemistry in the medical faculty at the University of Gottingen where he worked till he died in 1882.

Wöhler's years at the industrial school at Berlin between 1825 and 1831 were scientifically perhaps the most prolific years of his research career. During this period, he published his famous researches on aluminum, beryllium, yttrium, and silicon as well as cyanogen, cyanic and uric acids (in association with Liebig). His two outstanding scientific achievements are the isolation of aluminum and more famously, synthesis of urea.

Synthesis of urea and the birth of organic chemistry

Most scientists of this time including Berzelius believed that only living things could make "organic" compounds. Accordingly it was thought urea, a compound found in the urine of animals and humans, could be produced in the organs of living systems. This belief was based on the vitalism hypothesis which believed that a special "vital force" was responsible for living things being alive.

> Berzelius had proclaimed in his text-book that scientists would never be able to prepare the products of vital force in laboratories and that scientists could prepare as well as destroy material things, but to imitate the vital forces of organic origin would be impossible.

The synthesis of urea in the laboratory marked not only the beginning of the end of vitalism but more importantly the beginning of organic chemistry.

Urea is perhaps one of the most important and widely used compounds today – it is an important chemical fertilizer as well as an important industrial raw material.

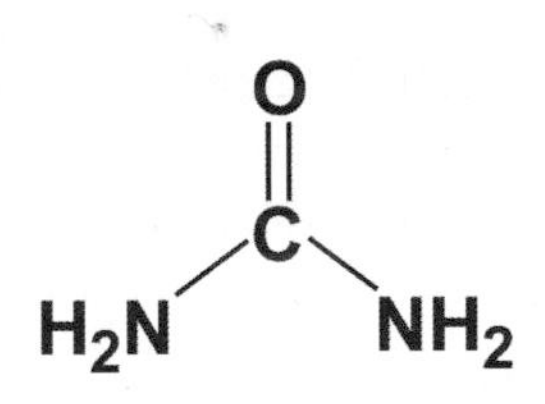

Wöhler had prepared urea in the laboratory from ammonium cyanate without employing even a single living cell. The importance of the synthesis of urea can be gauged by the fact that many scientists believe that organic chemistry owes its origin to this chemical.

Wöhler said *"This investigation has yielded an unanticipated result that reaction of cyanic acid with ammonia gives urea, a noteworthy result in as much as it provides an example of the artificial production of an organic, indeed a so-called animal, substance from inorganic substances"* (From J. C. Poggendorff [3]).

Wöhler's synthesis of urea, like many scientific discoveries, was the result of serendipity. Wöhler began his experiments to investigate cyanates, salts of cyanic acid in 1823. This may be regarded as the beginning of the accidental synthesis of urea. Wöhler tried to synthesize ammonium cyanate by reacting silver cyanate with aqueous ammonium chloride in 1828. He found that the properties of the resultant white crystalline solid product were different from the known properties of ammonium cyanate. Wöhler next attempted to synthesize ammonium cyanate by reacting lead cyanate with ammonium hydroxide. He obtained the same white powder as the product but with one difference - it was less contaminated. On analysing the white powder, Wöhler discovered that its composition and properties were similar to those of urea isolated from an organic source (urine).

$$Pb(OCN)_2 + 2NH_3 + 2H_2O \longrightarrow Pb(OH)_2 + 2NH_4OCN$$

$$\downarrow$$

$$2H_2NCONH_2$$

Ammonium cyanate formed in the reaction was an unstable compound. Consequently it rearranged into the more stable compound urea. Ammonium cyanate, an inorganic compound, had become urea an organic compound!

$$NH_4OCN \longrightarrow NH_3 + HNCO \longleftrightarrow H_2NCONH_2$$

Wöhler realized immediately that he had prepared an organic compound in the laboratory contrary to the belief prevalent at that time that organic compounds could be synthesized only within a living system. He wrote to Berzelius in a letter on 22 February 1828 *"I can no longer, as it were, hold back my chemical urine; and I have to let out that I can make urea without needing a kidney, whether of man or dog: the ammonium salt of cyanic acid is urea"* (From *Encyclopedia.com* [4]). Interestingly, initially Wöhler appeared unhappy by his achievement as expressed in his letter to Berzelius. He wrote *"The great tragedy of science, the slaying of a beautiful hypothesis (vitalism) by an ugly fact (synthesis of urea in his laboratory)"* (From the Webpage 'The human touch..... organic Chemistry' [5]).

Wöhler's letter to Berzelius dated 22 February, 1828

"Perhaps you can remember the experiments that I performed in those happy days when I was still working with you, when I found that whenever one tried to combine cyanic acid with ammonia a white crystalline solid appeared that behaved like neither cyanic acid nor ammonia. . . . I took this up again as a subject that would fit into a short time interval, a small undertaking that would quickly be completed and – thank God – would not require a single weighing.

The supposed ammonium cyanate was easily obtained by reacting lead cyanate with ammonia solution . . . Four-sided right-angled prisms, beautifully crystalline, were obtained. When these were treated with acids, no cyanic acid was liberated, and with alkali, no trace of ammonia. But with nitric acid lustrous flakes of an easily crystallized compound, strongly acid in character, were formed; I was disposed to accept this as a new acid because when it was heated, neither nitric nor nitrous acid was evolved, but a great deal of ammonia. Then I found that if it were saturated with alkali, the so-called ammonium cyanate reappeared; and this could be extracted with alcohol. Now, quite suddenly I had it! All that was needed was to compare urea from urine with this urea from a cyanate" (From *Encyclopedia.com* [4]).

Wöhler announced the synthesis of urea in the journal *Annalen der Physik und Chemie* in 1828 (From "Wöhler Synthesis", Wikipedia [6]). Wöhler realised that the white powder he had obtained as a by-product when he reacted cyanogen and aqueous ammonia to get oxalic acid in 1823 was actually urea! (He referred to this work in his paper).

Many influential chemists did not believe that organic substances could be made in the laboratory without using any living organism. The spin-off of this synthesis was that many industrially useful organic chemicals which could only be obtained from organic sources (like urea from urine, citric acid from lemons and benzene from benzoin gum) could now be produced in large quantities using readily available and cheaper inorganic raw materials. The earliest industrial exploitation of this discovery was in the manufacture of synthetic dyes and drugs in large quantities. It was only after Herman Kolbe obtained acetic acid from carbon disulfide in his laboratory without any living organism, that the idea of synthesizing organic molecules in the laboratory from inorganic compounds was accepted.

While there is no record of Berzelius's immediate response to the game-changing discovery of Wöhler, he acknowledged the profound value of this discovery in an oblique manner in his letter to Wöhler dated January 22, 1831.

Even after the successful synthesis of urea starting from inorganic compounds, Wöhler continued to be interested in the analysis of minerals. Wöhler sent a sample of an unknown substance to Berzelius (with a question mark). Berzelius on analysis found the substance was actually vanadium oxide. He wrote back to Wöhler the following letter.

STOCKHOLM ***January 22, 1831***

"*. . . . In reference to the specimen sent by you, designated with an interrogation-mark, I will relate the following story: In the* *remote regions of the north there dwells the goddess Vanadis, beautiful and lovely. One day there was a knock at her door. The goddess was weary, and thought she would wait to see if the knock would be repeated, but there was no repetition, and whoever it was went away. The goddess, curious to see who it could be to whom, it appeared to be a matter of so much indifference whether he was admitted or not, ran to the window to look at the retreating figure. "Ah!" said she to herself, "it is that fellow Wöhler; he deserves his fate for the indifference he showed about coming in." A short time afterward there was another knock at the door, but this time so persistent and energetic that the goddess went herself to open it. It was Sefström who appeared at the threshold, and thus it was that he discovered vanadium. Your specimen is, in fact, oxide of vanadium. But the chemist who has invented a way for the artificial production of an organic body can well afford to forego all claims to the discovery of a new metal, for it would be possible to discover ten unknown elements without the expenditure of so much genius as appertains to the masterly work which you, in association with Liebig, have accomplished and have just communicated to the scientific world"* (From Joy [2]).

It was the collaborative work on cyanates with Liebig, a life-long friend, that convinced Wöhler that it was indeed possible to prepare a variety of organic compounds in the laboratory from inorganic raw materials. In 1838, Liebig and Wöhler published in *Justus Liebigs Annalen der Chemie*, the results of their research in a long paper titled Untersuchungen über die Natur der Harnsäure (From Wöhler and Liebig [7]) in which they described (with the detailed analysis) 14 new compounds that they had prepared. Encouraged by Liebig, Wöhler wrote to Berzelius in 1838 *"The philosophy of chemistry will conclude from this work that it must be held not only as probable but [as] certain that all organic substances, in so far as they no longer belong to an organism, will be prepared in the laboratory. Sugar, salicin, morphine will be produced artificially. It is true that the route to these and products is not yet clear to us, because the intermediaries from which these materials develop are still unknown, but we shall learn to know them"* (From *Encyclopedia.com* [4]).

Even though successful synthesis of urea in the laboratory is recognized as the beginning of organic chemistry as an important branch of chemistry, Wöhler himself had grave misgivings about this new branch of chemistry. He wrote to Berzelius *"Organic chemistry just now is enough to drive one mad. It gives me the impression of a primeval, tropical forest full of the most remarkable things, a monstrous and boundless thicket, with no way of escape, into which one may well dread to enter"* (From Wöhler [8]).

Wöhler was also an accomplished inorganic chemist. His childhood fascination for minerals became a serious study of metals and preparation of mineral salts. Inorganic chemists of that period were merely collecting data. Wöhler introduced new methods of preparing useful inorganic compounds (for example heating sand with calcined bones to prepare phosphorus) and prepared new salts of a variety of metals. Many of his methods for preparing mineral salts have found industrial applications. He had the distinction of being the first chemist to get acetylene in 1862 from calcium carbide obtained by heating zinc, carbon and calcium.

Extraction of aluminum from alumina

A major achievement of Wöhler was the extraction of aluminum metal from alumina. Davy was the first chemist to try to isolate aluminum from alumina (wrongly listed as an element by Lavoisier in his Traite of 1789) in 1807. Many chemists including Berzelius had tried unsuccessfully to extract the metal from alumina. Oersted had come closest to achieving success in 1825 (he succeeded in getting an impure amalgam but could isolate the pure metal) but he neither published his preparation of aluminum chloride nor continued this work. He gave permission to his friend Wöhler, to continue experiments to extract aluminum. Wöhler used a completely new approach and succeeded in 1827 in extracting pure aluminum metal by heating in a platinum crucible a large quantity of anhydrous aluminum chloride and a small quantity of potassium. Wöhler published his findings in a paper titled *Über das Aluminium,* in *Annalen der Physik und Chemie*, 2nd ser., 11 (1827) describing the chemical properties such as its reaction with alkalies, acids as well as with other metals. He continued to work on aluminum and established in 1845 its important properties. He also succeeded in isolating beryllium (by heating beryllium chloride with potassium) and yttrium (by heating yttrium chloride with potassium) in 1828.

Wöhler's novel technique

Wöhler took a large quantity of aluminum chloride and a small amount of potassium in a covered platinum crucible. He then gently heated the crucible containing the mixture, it started to react vigorously. When he added water to the product of the reaction (metallic residue), no alkali was produced indicating the absence of potassium. Wöhler's experiment proved that metal thus obtained was not an alkali metal and went on to describe its chemical properties like its reactions with other metals, alkalies and acids. Later in the century, Napoleon III honoured him for the discovery of this important metal.

Honours

Wöhler was honoured by many countries for his outstanding contributions to science. Some of them are: Foreign member of The Royal Society, London (1854), Permanent Secretary, the Royal Hanoverian Academy of Sciences (1860), Foreign Associate of the French Academy (1864), an officer of the Legion of Honour, and Copley medal of The Royal Society, London (1872).

Gottingen, where he was a professor of chemistry at the Max Planck Institute, has a statue of its favourite son at the city centre.

Other Facets

In addition to being an outstanding chemist, Wöhler was also professor of pharmacy and director of many laboratories. As inspector general of apothecaries of Hannover kingdom, from 1836 to 1848, he toured extensively inspecting apothecary shops. In addition, he contributed to the growth of the school of chemistry at Gottingen and never shirked his responsibilities as a teacher during the forty six years he spent at Gottingen.

According to his own estimate, over 6,500 students attended his lectures between 1845 and 1866. He wrote text books of inorganic chemistry and organic chemistry which were translated into many European languages. He translated Berzelius's texts into German.

According to August Wilhelm Hofmann, Wöhler was *"...unimpassioned, meeting even the most malignant provocation with an immovable equanimity, disarming the bitterest opponent by the sobriety of his speech, a firm enemy to strife and contention...."* (From Hunter [9]).

Wöhler was loved as a human being, admired as a teacher and was respected for his encyclopedic knowledge of the subject. Wöhler had a good sense of humour which he used to good effect. He is remembered as a true scientist who remained interested in what was happening in his laboratory till his final days. He continued to work even when he was over seventy years old writing research papers, guiding and mentoring students and of course teaching. He breathed his last at the age of 82 on September 23, 1882.

References

1. Scientific American Supplement No. 362, New York, 9 December 1882.

2. Charles. A. Joy, Biographical Sketch of Frederick Wöhler, *The Popular Science Monthly*, Vol.17, (1880).

3. J. C. Poggendorff, *'On the Artificial Formation of Urea'. Annalen der Physik und Chemie*, Vol. 88, 253-256, 1828.

4. "Friedrich Wöhler", Complete Dictionary of Scientific Biography. 2008. *Encyclopedia.com.* 16 Jan. 2015.

5. From the Webpage, The human touch of Chemistry, History & Future - Urea and the beginnings of organic chemistry.

6. (*Annalen der Physik und Chemie* the paper "Ueber künstliche Bildung des Harnstoffs". *Annalen der Physik und Chemie* 88 (2): 253–256). (From Wikipedia, The Free Encyclopedia, Wöhler Synthesis).

7. Untersuchungen über die Natur der Harnsäure," in *Justus Liebigs Annalen der Chemie*, **26** (1838), 241–340, Wöhler and Liebig.

8. From Friedrich Wöhler's Letter to J. J. Berzelius (28 Jan 1835). From *Bulletin of the Atomic Scientists*, Nov. 1949.

9. Graeme K. Hunter, *Vital Forces: The Discovery of the Molecular Basis of Life* (San Diego: Academic Press, 2000), 35.

'Friedrich Wöhler', Source: http://www.sil.si.edu/digitalcollections/hst/scientific-identity/explore.htm, (From Wikipedia, The Free Encyclopedia).

7. AUGUST KEKULE (1829–1896)

First to predict organic structures

"Let us learn to dream, gentlemen, then perhaps we shall find the truth... But let us beware of publishing our dreams before they have been put to the proof by the waking understanding" (From Knight [1]).

Early life

(From Wikimedia Commons)

Kekule was born on September 7, 1829 in Darmstadt, Germany. Kekule could trace his ancestry to Czech nobility. One branch of the family migrated to Germany many centuries earlier and settled down in Darmstadt. His father, Ludwig Carl Emil Kekule, was a civil servant and young Kekule had a comfortable childhood. He started his education at the gymnasium in Darmstadt where apart from science subjects, he showed an aptitude for languages and drawing. After finishing at the gymnasium, he joined the University of Giessen in the winter semester of the academic year 1847–1848 to study architecture.

The turning point in Kekule's destiny came about when he took Liebig's chemistry course during the second semester. He was so impressed by

Liebig's lectures that he decided to change to chemistry. Kekule's father had died by then so the family council had to give Kekule permission to change the subject. After much persuasion, they agreed with Kekule's new plan to study chemistry instead of architecture. When he began to work with Liebig in 1850, Liebig warned Kekule *"If you want to become a chemist, you will have to ruin your health. If you don't ruin your health studying, you won't accomplish anything these days in chemistry"* (From Brock [2]).

As Liebig was preoccupied with collating material for his *Chemische Briefe,* he made Kekule responsible for the research on the chemical composition of wheat bran and gluten. He was so pleased with Kekule's results that he offered Kekule a position in his laboratory. However, Kekule refused the offer as he was more interested in theoretical aspects of chemistry. Strictly following the laboratory protocol did not suit his nature.

"Originally a pupil of Liebig, I became a pupil of Dumas, Gerhardt and Williamson; I no longer belonged to any school" (From Knight [1]).

Liebig accepted Kekule's decision and encouraged him to pursue his studies in Paris in 1851 under Charles Gerhardt who was known then for his unitary theory of chemistry, work on radicals and most of all, for systematization of organic compounds into four categories. After obtaining his doctorate in 1852, Kekule went to Switzerland for a year (1852-1853) and on Liebig's advice from there to London to work at St. Bartholomew's Hospital. Here he came under the influence of Williamson who was instrumental in Kekule developing an interest in theoretical chemistry. Williamson persuaded Kekule to work on further extending the work done by Gerhardt. Kekule was successful in adding a new type – the marsh gas or methane type. This work had an unexpected result and led to Kekule's discovery later that carbon was tetravalent.

Gambolling atoms and theory of Valence

While at London, Kekule began to work on radicals and on certain organic molecules. One evening while he was returning home by bus after visiting a colleague, Kekule reported that he had a vision of atoms "gambolling".

*"Whenever, hitherto, these diminutive beings (*atoms*) had appeared to me, they had always been in motion; but up to that time I had never been able to discern the nature of their motion. Now, however, I saw how, frequently, two smaller atoms united to form a pair, how the larger one embraced the two smaller ones; how still larger ones kept hold of three or even four of the smaller; whilst the whole kept whirling in a giddy dance. I saw how the larger ones formed a chain, dragging the smaller ones after them but only at the ends of the chain. The cry of the conductor 'Clapham Road', awakened me from my dreaming; but I spent part of the night in putting on paper at least sketches of these dream forms. This was the origin of the Structurtheorie (Structural Theory)"* (From Knight [1]). This mental picture, helped perhaps by his interest in architecture gave Kekule the idea of carbon atoms joining to form chains. From this fantasy, Kekule's theory of valence was born.

"*Carbon is, as may easily be shown and as I shall explain in greater detail later, tetrabasic or tetratomic, that is 1 atom of carbon, (C = 12) is equivalent to 4 atoms of H*" (From Benfey [3]).

Kekule began his teaching career in the summer of 1856 at the University of Heidelberg where he taught organic chemistry. He was involved at the same time in doing research to understand the structure of carbon compounds. By conducting many experiments on carbon compounds containing a single carbon atom as in CH_4, CCl_4, CH_3Cl, CO_2, he came to the conclusion that *"when the simplest compounds of this element are considered (marsh gas, chloride of carbon, chloroform, carbonic acid, phosgene, sulphide of carbon, hydrocyanic acid, etc.) it is seen that the quantity of carbon which chemists have recognised as the smallest possible, that is, as an atom, always unites with 4 atoms of a monatomic or with two atoms of a diatomic element; that in general, the sum of the chemical units of the elements united with one atom of carbon is 4. This leads us to the view that carbon is tetratomic or tetrabasic"* (From Partington [4]). After establishing that carbon was tetravalent in simple compounds, he continued his research into carbon compounds containing more than one carbon atom. He stated that in compounds having more than one carbon atom, part of the affinities of the carbon atom was satisfied by the other carbon atoms present and the rest by the affinities of other elements or radicals present in it. The process continues even when there are more than two carbon atoms forming an open carbon chain.

Kekule used old fashioned shadowed figures drawn on a blackboard and paper cuttings as his favourite teaching aids to clarify graphically this concept in his lectures. This imaginative way of teaching a completely new theory made his courses popular, particularly his diagrams of molecular structure. Before his discovery, scientists believed that molecule had only one central atom to which all other atoms of the molecule were bound. While a number of chemists welcomed Kekule's theory as it explained the reason for the staggering number of carbon compounds in existence, some chemists, particularly Kolbe did not accept it. He wanted physical evidence, but alas X-ray crystallography and such other techniques were not available at that time.

Kekule in Germany and Archibald Scott Couper in Scotland both arrived independently and simultaneously at the same conclusion that carbon atoms had the capacity to link with other carbon atoms and thus form a chain. It is interesting that each had a definite contribution to make in this historic discovery. Kekule's contribution was his discovery in 1857 that carbon was tetravalent and Couper's was the innovative use of straight lines and chemical symbols of elements to represent the linking of the carbon atoms with other elements.

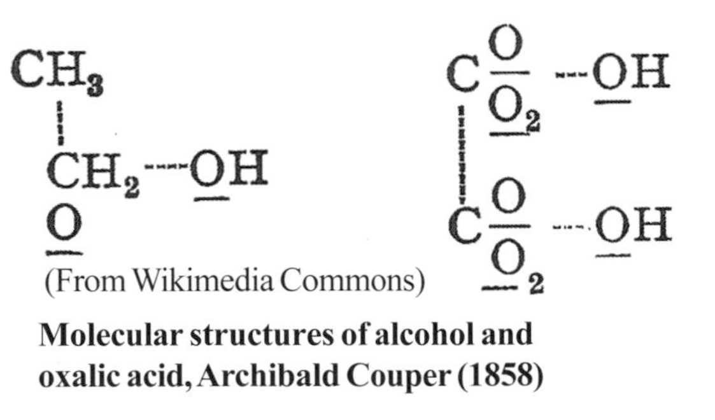

(From Wikimedia Commons)

Molecular structures of alcohol and oxalic acid, Archibald Couper (1858)

Kekule got the credit for proposing this concept as his paper was published in Liebig's *Annalen der Chemie und Pharmacie* in May 1858 three weeks earlier than Couper's paper which was published later due to a delay because of his superior Charles Adolphe Wurtz, in whose laboratory in Paris he was working at that time. Wurtz delayed presenting the paper to the French Academy. Couper felt cheated of legitimate recognition as his paper contained more examples through solved formulae than Kekule's paper. He had even suggested that it was possible to have cyclical formulae. This suggestion may have given a clue to Kekule in his later work on the structure of benzene. This disappointment and the ensuing bitter fight with Wurtz (Wurtz threw him out of his institute) had a disastrous effect on Couper. He had a series of nervous breakdowns and gave up research completely. He spent the rest of his life in and out of hospitals and was cared for by his mother. However, even though the credit for the theory of the capacity of carbon atoms to link with each other and form a chain was given only to Kekule, credit for developing the structural theory of organic chemistry belongs to the efforts of both Kekule and Couper.

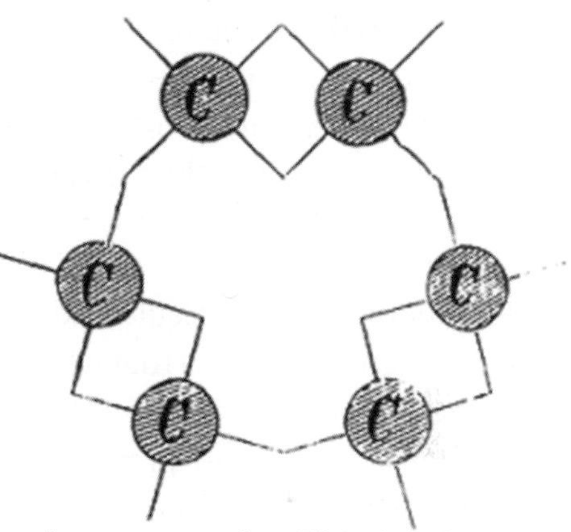

A representation Kekule's benzene ring from *Lehrbuch der organischen Chemie* (1861–1867). CHF Collections.

Benzene

"I would trade all my experimental work for the single idea of the benzene theory" - August Wilhelm von Hofmann (From Lepsius[5]).

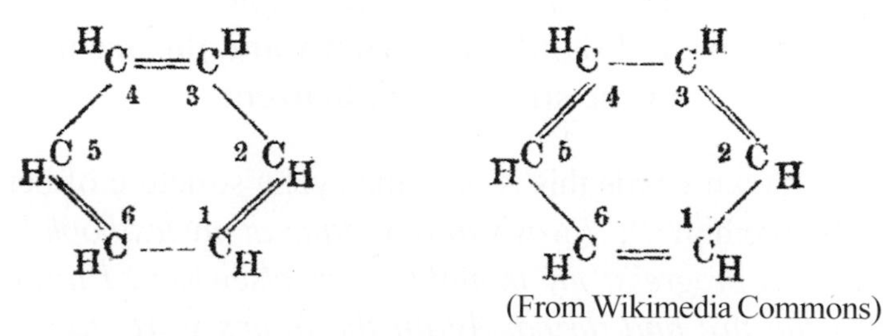

(From Wikimedia Commons)

August Kekule's proposal for the benzene structure

In 1858, Kekule accepted the attractive offer of full professorship from the University of Ghent, in the French speaking part of Belgium. His stay at the University is marked by outstanding contributions to the chemistry of unsaturated aromatic carbon compounds. It was here Kekule made the discovery of the structure of benzene which is considered by many as one of the greatest discoveries. The structure of benzene was a challenge to chemists all over Europe as benzene, an unsaturated aromatic carbon compound, had a structure that was unlike the linear structure of the known carbon compounds.

Kekule initially found it difficult to adjust to an unfamiliar environment and teaching in French. His daily routine consisted of working in the laboratory the entire day, working on the first part of his text book *Lehrbuch der organischen Chemie* and preparing for the next day's lecture after midnight! In addition, he also organized the first International Congress of Chemists in 1860 to arrive at chemical terminology acceptable to chemists everywhere.

As Kekule had considered the tetravalence of carbon to be fixed, he had to explain how the valence of carbon was satisfied in unsaturated carbon compounds. He had already published in 1862 his model to explain the presence of double bonds between two carbon atoms next to each other in unsaturated isomers. The problem of the structure of benzene became the most important problem to solve. It was known that benzene consisted of six carbon and six hydrogen atoms. However, all the known structural formulae (including Kekule's own open chain structure) could neither be applied to benzene nor explain its unique properties.

The dream of Ouroboros, the symbol of snake seizing its own tail signifying cyclicity and the prediction of the structure of benzene

In Kekule's own words this is how the cyclic structure of benzene flashed before him *"...I was sitting writing at my textbook but the work did not progress; my thoughts were elsewhere. I turned my chair to the fire and dozed. Again the atoms were gambolling before my eyes. This time the smaller groups kept modestly in the background. My mental eye, rendered more acute by the repeated visions of the kind, could now distinguish larger structures of manifold confirmation: long rows, sometimes more closely fitted together; all twining and twisting in snake-like motion. But look! What was that? One of the snakes had seized hold of its own tail, and the form whirled mockingly before my eyes. As if by a flash of lightning I awoke; and this time also I spent the rest of the night in working out the consequences of the hypothesis"* (From Knight [1]). The snake catching it's own tail gave Kekule the idea that it was possible to arrange the six carbon atoms of benzene as a closed ring or circular structure if there were three double bonds between three carbon atoms and three single bonds between the remaining three carbon atoms and if the double and single bond atoms were in dynamic equilibrium. This provided a satisfactory solution to the vexing problem of benzene structure. He compared the wriggling motion of the snakes of his dream to the oscillations of the shared valence electrons.

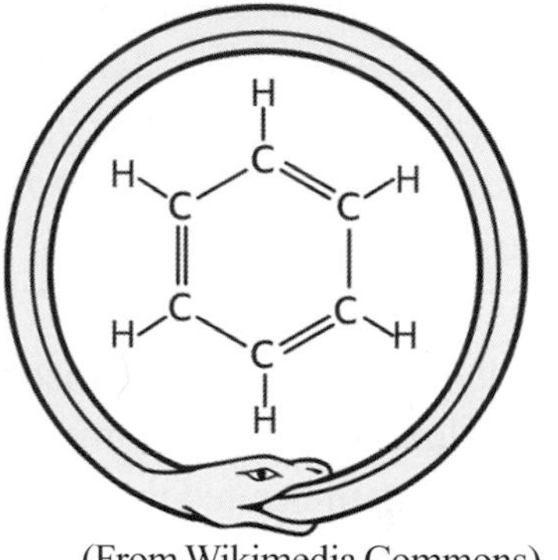

(From Wikimedia Commons)

Was Kekule a bee keeper?

In an after-dinner speech at Kekule Benzolfest in Mar 1890, August Wilhelm von Hofmann said *"People have wracked their brains for an explanation of benzene and how the celebrated man [Kekule] managed to come up with the concept of the benzene theory. With regard to the last point especially, a friend of mine who is a farmer and has a lively interest in chemistry has asked me a question which I would like to share with you. My 'agricultural friend' apparently believes he has traced the origins of the benzene theory. 'Has Kekule,' so ran the question, 'once been a bee-keeper? You certainly know that bees too build hexagons; they know well that they can store the greatest amount of honey that way with the least amount of wax. I always liked it,' my agricultural friend went on, "'When I received a new issue of the Berichte; admittedly, I don't read the articles, but I like the pictures very much. The patterns of benzene, naphthalene and especially anthracene are indeed wonderful. When I look at the pictures I always have to think of the honeycombs of my bee hives"* (From Brock, Benfey & Stark [6]).

Kekule had arrived at his theory of the cyclic structure of benzene in 1862. However, it was only in 1865 that his theory was made public when Wurtz presented the paper on benzene to the *Société chimique de Paris*. Later in the year, it was published as a paper titled "Sur la constitution des substances aromatiques" in the *Bulletin de la Société chimique de Paris* paper. Soon after this, it was published in German as a paper titled "Untersuchungen uber aromatische Verbindungen" and was presented to the Académie on 11 May 1865, in which using benzene's geometry Kekule tried to find out how many monosubstituted, disubstituted, and trisubstituted isomeric derivatives of benzene could be obtained. He and his co-workers then initiated experiments to get proof of their predictions. It was many years later they were able to obtain the experimental proof.

"In the benzene nucleus we have been given a soil out of which we can see with surprise the already-known realm of organic chemistry multiply, not once or twice but three, four, five or six times just like an equivalent number of trees. What an amount of work had suddenly become necessary, and how quickly were busy hands found to carry it out! .... Everywhere in the sea of leaves one can spy the slender hydroxyl bud: hardly rarer is the forked blossom [Gabelblüte] which we call the amine group, the most frequent is the beautiful cross-shaped blossom we call the methyl group. And inside this embellishment of blossoms, what a richness of fruit, some of them shining in a wonderful blaze of color, others giving off an overwhelming fragrance" - August Wilhelm von Hofmann (From Brock, Benfey & Stark [6]).

The prediction of the closed ring structure of benzene was hailed by scientists and Arthur Koestler in his book The Act of Creation, (1964) termed it as *"the most brilliant piece of prediction to be found in the whole range of organic chemistry ....and one of the cornerstones of modern science"* (From Koestler [7]). Professor Japp in the Kekule memorial lecture delivered at the London Chemical Society meeting in December 1897 said *"Kekule's benzene theory was the most brilliant piece of scientific prediction to be found in the whole of organic chemistry.... three fourths of modern organic chemistry is directly or indirectly the product of his theory"* (From Knight [1]).

The chemistry of benzene did not conform with this structural formula and Kekule modified his theory in 1872 to overcome this problem. He suggested that there were two forms of benzene and the bonds were constantly changing their position in the two isomers. In other words, this theory assumed that since each of the six carbon-carbon bond in benzene is a single bond half the time and double bond the other half of the time, all the carbon-carbon bonds are equivalent. Even the modified theory of Kekule was only partially correct and it was left to Linus Pauling almost 68 years later (in 1933) to explain satisfactorily the structure of benzene by using quantum mechanics. He used the concept of resonance to explain the structure of benzene instead of Kekule's theory of oscillating bonds.

Return to Germany as professor in the University of Bonn

When he was offered the chair of chemistry at the University of Bonn in 1867 with the promise of a new Institute devoted to chemistry, Kekule gladly accepted it and returned to Germany in September of that year. He soon established an outstanding chemistry laboratory there and many leading chemists including van't Hoff came to listen to his lectures. Kekule was committed to educational reforms to advance the principles of higher education. He also resumed writing his textbook. His original plan for the remaining two volumes had to be curtailed drastically as the scope of chemistry had changed beyond recognition.

1890 marked another major landmark in his professional life when he presented the paper 'Ueber die Konstitutionen des Pyridins' in which he summed up the research findings of his investigations of pyridine from 1886 at the general assembly meeting of Deutsche Chemische Gesellschaft held in Berlin on March 10th. It was received with enthusiasm as its formula could be compared with that of benzene.

Honors

Kekule received The Copley Medal of the Royal Society of London in 1885. In 1895 he was made a lifetime peer and took the name August Kekule von Stradonitz.

Last years

Kekule died before Noble Prizes were instituted. His great influence on chemistry can be judged by the fact that between 1901 and 1905 three of his students received Nobel Prizes in Chemistry (1901 - van't Hoff first Nobel Prize in Chemistry, 1902 - Emil Fischer, first Nobel Prize in Organic Chemistry and 1905 - Baeyer).

Kekule continued to do research practically till the end of his life. After a brief illness, Kekule died in Bonn, on 13 July 1896 at a relatively young age of 66.

August Kekule had an indelible impact on organic chemistry. Even today his pioneering contributions are honoured by chemists of all hues.

References

1. David M. Knight (ed.), *The Development of Chemistry, 1789-1914: Selected essays,* Routledge, 1998.
2. William H. Brock, *Justus Von Liebig: The Chemical Gatekeeper*, Cambridge University Press, 1997. (As quoted by R.E. Oesper, *The Human Side of Scientists*, (Cincinnati, 1975, p 108).

3. Otto Theodor Benfey, *From Vital Force to Structural Formulas,* First Reprint Edition, 1975, published by the American Chemical Society. p74.
4. J. R. Partington, *A History of Chemistry* (1972), Vol. 4, 536.
5. Quoted by B. L. Lepsius in 'Hofmann und die Deutsche Chemische Gesellschaft', Berichte der Deutschen Chemischen Gesellschaft (1918), 51, (Benzene, (2014, October 31). Wikiquote, Retrieved March 23, 2015).
6. William H. Brock, O. Theodor Benfey and Susanne Stark, '*Hofmann's Benzene tree at the Kekule festivities*', J. Chem. Educ., 1991, 68 (11), p 887-888.
7. Arthur Koestler, *The Act of Creation,* 1964, p118.

August Kekule, 1880 (Wikimedia Commons). (http://sceti.library.upenn.edu/sceti/smith/scientist.cfm?PictureID=2235&ScientistID=15)

"Ouroboros-benzene" by Haltopub (Licensed under CC BY-SA 3.0 via Wikimedia Commons - http://commons.wikimedia.org/wiki/File:Ouroboros-benzene.svg#/media/File:Ouroboros-benzene.svg)

A representation of the benzene ring from Kekule's *Lehrbuch der organischen Chemie* (1861–1867). CHF Collcctions.

Archibald Scott Couper, 1858 upload: 31 December 2006, Source: Transferred from en.wikipedia; transferred to Commons by User:Hystrix using CommonsHelper (Wikimedia Commons).

August Kekule's proposal for the benzene structure in its original, taken from *August Kekule (1872). "Ueber einige Condensationsproducte des Aldehyds". Liebigs Ann. Chem.* ***162*** *(1): 77–124., 16 August 2010,* Kekule_-_Ueber_einige_Condensationsproducte_des_Aldehyds.pdf - source (From Wikimedia Commons).

8. DMITRI IVANOVICH MENDELEEV (1834–1907)

Designer of the greatest table

"For the first time, I saw a medley of haphazard facts fall into line and order. All the jumbles and recipes and hotchpotch of the inorganic chemistry of my boyhood seemed to fit themselves into the scheme before my eyes - as though one were standing beside a jungle and it suddenly transformed itself into a Dutch garden"- C. P. Snow (From Snow [1]).

(From "Dmitri Mendeleev", Wikipedia)

The Periodic Table is considered as the greatest storehouse of chemical information about known elements. Though the modern periodic table is the end result of the efforts of many scientists, the credit for designing it to give a fixed place to the then known elements belongs undoubtedly to Dmitri Mendeleev of Russia. Today, a discerning user can extract enough information from the table to create new compounds and materials.

Early life

Dmitri Ivanovich Mendeleev was born to Ivan Pavlovich Mendeleev and Maria Dmitrievna Kornilieva on February 7, 1834 in Verkhnie Aremzyani, a small village in Siberia close to Tobolsk. He was the youngest in a large family. His father taught politics, philosophy and fine arts in the local gymnasium (high school). His mother's family introduced paper and glass making in Siberia after settling down there in the early 1700s. The large family had to face hard times as his father lost his job after becoming blind and succumbed to tuberculosis shortly after. The burden of providing for a large family now fell on his mother's shoulders. Maria's family came to her rescue by allowing her to manage the family owned glass factory in Aremziansk so that she could supplement her meager widow's pension. Maria moved to Aremziansk with her children.

Maria was Dmitri's first and most influential teacher. As she had to work to feed the family, Maria taught her young son correcting him gently and instilling in him human values and love for truth. Even as a young boy, Dmitri could understand complex concepts and his mother was convinced that Dmitri was exceptionally talented. She was determined that he should attend university when he grew up and devote his life to the pursuit of science. She started saving money towards this end. As Dmitri was still young, his mother allowed him to spend time in the glass factory where he spent his time interacting with the workers. The glass factory was also an informal school as he learnt about the intricacies of glass making from the chemist in the factory and the art of making beautiful glass objects from the glass blower. Dmitri was also lucky to have his brother-in-law Bessargin teaching him science.

Three mantras Dmitri learnt early in life

"Everything in the world is love" from his mother Maria.

"Everything in the world is art" from Timofei, the glass blower at the glass factory.

"Everything in the world is science" from Bessargin, his brother–in-law. (From Posin [2])

Dmitri enrolled in the gymnasium in Tobolsk when he was 14 years old. Just as he was settling down, another major tragedy struck the family – a fire in his mother's glass factory destroyed it completely. The family had no resources to rebuild the factory, the only money the family had was what Maria had so diligently saved for Dmitri's education. Maria was undeterred by this and was determined that Dmitri should get university education. She realized that the only way to do this was for him to win a scholarship to attend the university. Maria pushed Dmitri towards this goal. In those days it was necessary to perform well in classical languages and subjects to win the scholarship. As Dmitri had decided to study science, he considered it a waste of his time to study classical subjects. He was convinced that these subjects would not help in his future career as a scientist. This attitude towards what he considered wasteful education remained throughout his life. He was convinced that what meaningful education required was not Plato but many more Newtons. He declared in *1901 "We could live at the present day without a Plato, but a double number of Newtons is required to discover the secrets of nature, and to bring life into harmony with the laws of nature"* (From Harrow [3]).

Within a year of his entering the gymnasium at Tobolsk, his mother Maria decided to leave Aremziansk in Siberia, to settle down in Moscow to give young Dmitri a better chance of university education. It is said that Dmitri, his sister and his mother walked almost 1,600 kilometers all the way to Moscow as they were too poor to afford any transportation. There was great political upheaval in Moscow and Dmitri as an outsider could not get admission. Dmitri was upset by this rejection but not Maria. Such was Maria's determination that she decided to move to St. Petersburg with her two children to try for better luck. Again the three of them walked from Moscow to St. Petersburg. Even though they faced similar difficulties at St. Petersburg, Dmitri was able to take the entrance examinations of the Pedagogical Institute. On the basis of adequate performance in the examinations, Dmitri was admitted to the Institute in 1850 on full scholarship to the science teacher training programme. At last, Maria's dream was fulfilled. Maria had spent not only the last of her meager savings but her physical strength. Soon after he began his work at the Institute, Dmitri was dealt a double blow emotionally. First his mother died seven days after he joined the course and shortly after, his sister passed away, both victims of tuberculosis. For the first time in his life, Dmitri was left alone to face the challenges of managing his life and work at the Institute. Dmitri's total devotion to his mother and his acknowledgement of her sacrifices so that he could fulfill his scientific destiny is brought out poignantly in the dedication of his dissertation to his mother: *"This investigation is dedicated to the memory of a mother by her youngest offspring.....conducting a factory, she educated me by her own word, she instructed by example, corrected with love, and in order to devote him to science she left Siberia with me, spending thus her last resources and strength. When dying she said refrain from illusion; insist on work and not words. Patiently search divine and scientific truth"* (From Tilden [4]).

Despite a setback due to a serious illness, he graduated at the top of his class. He was only 21 years old when the doctors suspected tuberculosis and predicted that he had only two more years to live unless he shifted to a healthier climate. Dmitri was determined to live long enough to achieve his own and his mother's ambition. He therefore took up a job as chief science master of the gymnasium at Simferopol in the Crimean Peninsula near the Black Sea in 1855. The warm climate there helped him to recover completely from the threat of tuberculosis and he returned to St. Petersburg a year later in 1856 and defended successfully his Master's thesis *'Research and Theories on Expansion of Substances due to Heat'*.

> Passion for teaching and research and extraordinary love for his country and countrymen were the two driving forces in Mendeleev's life. His love for teaching and research resulted in his outstanding contributions to chemistry and of course the Periodic Table and his devotion to Russia and Russian people led to his interest in chemical technology and contributing to such diverse fields as Russian agriculture, mining industries, transport and metrology.

Later years

In 1859, Mendeleev got his lucky break to broaden his scientific horizon when he was deputed for two years to visit well known research establishments in France and Germany to study the scientific and technological innovations taking place there. He got the opportunity to work with Regnault on densities of gases in Paris and studied the workings of the spectroscope with Kirchoff in Heidelberg. The studies he conducted on surface tension and capillarity helped to formulate the theory of absolute boiling point which later came to be known as critical temperature. Most importantly, he attended the first International Chemistry Congress held at Karlsruhe in 1860, where it was decided to adopt a universal system to measure the weights of different elements. It was also agreed that hydrogen's atomic weight be fixed as 1 as it was the lightest element. This decision contributed significantly to Mendeleev formulating the periodic table later.

Mendeleev returned to St. Petersburg in 1861 and almost immediately got immersed in teaching and research. He was appointed as the Professor of Chemistry at the Technological Institute in 1863 and the University of St. Petersburg made him a Professor of Chemistry in 1866. He transformed the University of St. Petersburg into an internationally acclaimed center for chemical research. He obtained his doctorate for his dissertation '*On the Combinations of Water with Alcohol*'. Much of his research was focused on finding solutions to problems facing his countrymen. He gave great importance to education. He was a dedicated teacher, equally interested in teaching students in formal classrooms and talking to poor farmers in the third class compartments of trains whenever he travelled. He was adored both by the university students and peasants. Just as the classrooms would be overflowing with students to hear his chemistry lectures, peasants would surround him when he discussed about the problems of farmers.

Mendeleev began publishing his research findings from 1854 onwards. '*Chemical Analysis of a Sample from Finland*' was his first publication. Mendeleev had over 250 publications and the list included his well known books, *Organic Chemistry* and *Principles of Chemistry. Organic Chemistry* was published in 1861 when Mendeleev was 27 years old.

Beginnings of the Periodic Table

It all started with Mendeleev's teaching at the University of St. Petersburg. To his dismay, he did not find any chemistry textbook that explained concepts clearly. He decided to write one himself. His periodic table was an important part of the book "Principles of Chemistry". From the time he began to work in chemistry, he was convinced that there was some kind of inherent order to the elements and he had been collecting all available data for nearly thirteen years to test his belief. This deep conviction about the existence of a "general reign of order" had a profound influence on his approach to chemistry and resulted in the creation of the Periodic Table.

From early 1800, many chemists, notably Newland from England, Chancourtois from France and Meyer from Germany had been working on the secret of periodicity. Each of them used a different yardstick to discover the basis of periodicity. For example while Newland discovered the periodicity of a few selected elements arranged on the basis of atomic weight, every eighth element showed similar properties. Newland called it the Law of Octaves. Chancourtois proposed that properties of elements are the properties of numbers. These scientists were grouping the known elements either on the basis of their common properties or by their atomic weights. Mendeleev was not aware of the work being done elsewhere as he was beginning to understand the nature of periodicity.

Mendeleev remarked *"...nothing, from mushrooms to a scientific dependence can be discovered without looking and trying. So I began to look about and write down the elements with their atomic weights and typical properties, analogous elements and like atomic weights on separate cards, and soon this convinced me that the properties of the elements are in periodic dependence upon their atomic weights; If all the elements are arranged in the order of their atomic weights, a periodic repetition of properties is obtained. This is expressed by the law of periodicity"* (From Mendeleev [5]).

Like the other chemists, Mendeleev also realized the importance of atomic weight in determining the properties of elements. In a paper that he published in 1869, he wrote *"No matter how properties of simple bodies may change in the free state, something remains constant........only one constant peculiar to an element .....(is) the atomic weight. Atomic weight belongs not to coal or diamond but carbon"* (From Kaji [6]).

In 1860, 56 elements were known. By 1863, another seven elements had been discovered and Mendeleev had to accommodate 63 elements. He methodically wrote the properties of elements and their atomic weights on individual cards. Mendeleev was a keen cards player. It is said that one day, while he was arranging the cards into suits horizontally and numbers vertically for playing patience (solitaire), it suddenly struck him that if he arranged the cards of elements in an ascending order of their atomic weights, certain type of elements occurred at regular intervals. He was convinced that he was close to unravelling the secret and constantly arranged and rearranged the cards many times. It is part of the folklore that in 1869 exhausted by the efforts, Mendeleev fell asleep at the table and later recounted *"In a dream I saw a table where all the elements fell into place as required. Awakening, I immediately wrote it down on a piece of paper"* (From Strathern [7]). When he began to classify and arrange the known elements according to their chemical properties based on their atomic weights, an inherent pattern became evident in the properties exhibited by the 56 elements only when he left blank places. He also observed elements with similar (chemical) properties appeared one below another in the vertical columns of the table. He declared *"If all the elements are arranged in the order of their atomic weights, a periodic repetition of properties is obtained. This is expressed by the law of periodicity"* (From Mendeleev [5]).

One of the factors that restricted Newland and others was that each of them used only the observed characteristics for classification and their tables had no provision for future discoveries. Mendeleev correlated all the available data such as periodicity, groups and chemical properties. Mendeleev realized that he could fit atomic weight, properties and his own observations to arrange the elements in the form of a table. He recognized that this was not merely a method of classifying the elements on the basis of certain observed facts, but was a "law of nature" which could be used to predict new information and also to show errors in what had been accepted as facts.

By 1869, Mendeleev had grouped all the known elements based on atomic weights and valences, giving detailed descriptions of each element. The elements were first arranged in rows (periods) in an ascending order of their atomic weights with each row having seven elements including blank places. Mendeleev noticed that the first element of each row had similar properties as the first element of its previous row! As hydrogen did not fit into the pattern, Mendeleev started the first row with lithium. He summed

ОПЫТЪ СИСТЕМЫ ЭЛЕМЕНТОВЪ,

ОСНОВАННОЙ НА ИХЪ АТОМНОМЪ ВѢСѢ И ХИМИЧЕСКОМЪ СХОДСТВѢ.

			Ti=50	Zr=90	?=180.
			V=51	Nb=94	Ta=182.
			Cr=52	Mo=96	W=186.
			Mn=55	Rh=104,4	Pt=197,1.
			Fe=56	Ru=104,4	Ir=198.
			Ni=Co=59	Pd=106,6	Os=199.
H=1			Cu=63,4	Ag=108	Hg=200.
	Be= 9,4	Mg=24	Zn=65,2	Cd=112	
	B=11	Al=27,3	?=68	Ur=116	Au=197?
	C=12	Si=28	?=70	Sn=118	
	N=14	P=31	As=75	Sb=122	Bi=210?
	O=16	S=32	Se=79,4	Te=128?	
	F=19	Cl=35,5	Br=80	I=127	
Li=7	Na=23	K=39	Rb=85,4	Cs=133	Tl=204.
		Ca=40	Sr=87,6	Ba=137	Pb=207.
		?=45	Ce=92		
		?Er=56	La=94		
		?Yt=60	Di=95		
		?In=75,6	Th=118?		

Д. Менделѣевъ

(From Wikimedia commons)

1869 version of Mendeleev's table. Notice the many gaps in the table

up his observation thus: Properties of elements vary periodically with the atomic mass or atomic weight. A formal presentation of Mendeleev's table titled *'The Dependence between the Properties and the Atomic Weights of the Elements'* was made to the Russian Chemical Society in 1869 by his colleague Professor Menshutken as Mendeleev was too ill to present it personally.

European scientists came to know of the discovery of the periodic law by Mendeleev and the publication of his Periodic Table of elements by a brief note published in *Zeitschrift fur Chemie* (*Journal of Chemistry*), in 1869. Mendeleev remarking about the importance of his Periodic Table, said *"When the elements are arranged in vertical columns according to increasing atomic weight, so that the horizontal lines contain analogous elements again according to increasing atomic weight, an arrangement results from which several general conclusions may be drawn"* (From Mendeleev [8]).

Mendeleev was 35 years old when the initial paper was presented. He was not satisfied with this table as he noticed that when the elements were arranged in the tabular form, he had to skip places to maintain similar chemical properties in the same vertical column or group. He was convinced that this was due to the missing elements – elements yet to be discovered. For example, when known elements were arranged, in the carbon group (group IV), he was convinced that tin could not occupy the space immediately below silicon. He left a gap for this yet to be discovered element. Similarly, in the group having elements boron, aluminum and yttrium (group III), he left two gaps below aluminum. He also predicted two more as yet undiscovered elements. He named the three elements yet to be discovered eka-silicon to fill the gap below silicon and eka-aluminum, eka-boron in the gaps below aluminum. On November 29, 1870 he declared that based on the periodicity of properties shown by known elements when arranged according to their atomic weights, it was possible to predict the properties of elements of these three elements. His contemporary chemists did not accept this revolutionary idea, some even ridiculed it. However, in 1875 French scientist Lecoq de Boisbaudran discovered the element predicted in the boron group and found the properties of the new element matched closely with Mendeleev's prediction of eka aluminum. In 1879, Lars Nilson of Scandinavia and Winkler of Germany in 1886 discovered the two elements predicted by Mendeleev for eka-boron and eka silicon. (It is said that Mendeleev used the term eka meaning one in Sanskrit as he had great respect for that language). The newly discovered elements fitted perfectly into the blank spaces provided in Mendeleev's Periodic Table. The new elements were named gallium for eka-aluminum (Gallia being the Latin name for France), scandium (for eka-boron) and germanium (for eka-silicon). He then summed up his observations and formulated his famous periodic law which states that the properties of elements vary periodically with the atomic mass. Mendeleev's prediction regarding the existence of these elements was accepted by the Society of Chemists in 1887. The vindication of his predictions and acceptance of his law of periodicity by his peers took him to the top of the scientific world.

As atom was accepted as the most basic constituent of matter at that time, Mendeleev and others had used atomic mass or atomic weight to classify the elements. It was known then that atomic number rather than atomic mass determined the chemical properties of elements. However, Mendeleev and others were able to determine the atomic weight of each of the sixty odd elements known at that time by comparing the atomic weight of the element to that of an atom of hydrogen as it was the lightest element known at that time.

Reihen	Gruppe I. — R^2O	Gruppe II. — RO	Gruppe III. — R^2O^3	Gruppe IV. RH^4 RO^2	Gruppe V. RH^3 R^2O^5	Gruppe VI. RH^2 RO^3	Gruppe VII. RH R^2O^7	Gruppe VIII. — RO^4
1	H=1							
2	Li=7	Be=9,4	B=11	C=12	N=14	O=16	F=19	
3	Na=23	Mg=24	Al=27,3	Si=28	P=31	S=32	Cl=35,5	
4	K=39	Ca=40	— =44	Ti=48	V=51	Cr=52	Mn=55	Fe=56, Co=59, Ni=59, Cu=63
5	(Cu=63)	Zn=65	— =68	— =72	As=75	Se=78	Br=80	
6	Rb=85	Sr=87	?Yt=88	Zr=90	Nb=94	Mo=96	— =100	Ru=104, Rh=104, Pd=106, Ag=108
7	(Ag=108)	Cd=112	In=113	Sn=118	Sb=122	Te=125	J=127	
8	Cs=133	Ba=137	?Di=138	?Ce=140	—	—	—	— — — —
9	(—)	—	—	—	—	—	—	
10	—	—	?Er=178	?La=180	Ta=182	W=184	—	Os=195, Ir=197, Pt=198, Au=199
11	(Au=199)	Hg=200	Tl=204	Pb=207	Bi=208	—	—	
12	—	—	—	Th=231	—	U=240	—	— — — —

(From wiki.chemprime.chemeddl.org)

Mendeleev's Periodic Table (1871)

What makes Mendeleev's periodic table unique?

Mendeleev created a table where all the known elements were arranged logically in a matrix. When one reads across the horizontal rows (periods) one set of related properties emerge and when read vertically (up or down) the groups reveal similar chemical properties - i.e. they reveal different sets of relationships across the periods and down the groups. Based on the position of an element – both its position in the period and in the group - it is easy to predict how the element will react with another element. It is an invaluable tool for research even today.

(From "Dmitri Mendeleev", Wikipedia)

The scientist's sculpture next to his Periodic Table on a wall of D. I. Mendeleyev Institute for Metrology in Saint Petersburg.

Mendeleev added more elements following the same criteria for classification and published an extended version some years later. Mendeleev did not agree with some of the accepted atomic weights at that time (the atomic weights could be measured only approximately at that time) as they did not correspond to the atomic weights suggested by his periodic law. For example, he noticed that tellurium had a higher atomic weight than iodine (even though he assigned the correct places for these elements but his reasoning was proved wrong). He also could not decide the position of the known lanthanides but again predicted that there was another row to his table which can accommodate some of the elements with the heaviest mass (actinides). There was no place for noble gases even in his later version of the periodic table. In spite of the many drawbacks in Mendeleev's table, it was used by chemists around the world for nearly 50 years. It laid the foundation for the Long Form of the Periodic Table in use today.

The many faceted genius

Even though Mendeleev is remembered mainly for his path breaking periodic table, he made other contributions to diverse fields that included chemistry, technology and agriculture as well as to industries. Mendeleev's important contributions include investigations into the expansion of liquids when heated, derivation of the formula with respect to the uniformity in the expansion of gases, (His formula was similar to Gay-Lussac's law), capillarity of liquids and surface tension which later led to his theory of 'absolute boiling point' (1861), properties and behavior of gases at high as well as low pressures, development of an accurate differential barometer. The absolute boiling point of a liquid was defined as the temperature at which both cohesion and the heat of vapourisation reach zero (irrespective of the pressure and volume) and at which temperature the liquid changes to vapour (1861). This anticipated the concept of the critical temperature of gases put forward by Thomas Andrews in 1861. Mendeleev also made significant contributions to the understanding of the nature of solutions.

"Dmitri was devoted to two things: First, his work and his students. Second, his country and his fellow men. His first love led him to write many books and to organize the periodic table, while the other gave rise to the studies of chemical technology and the organization of Russia's industries, agriculture, transport, meteorology, and metrology" (From Holmyard [9]).

Apart from his passion for teaching and research, he contributed to the technological advances of Russia. He did research in agricultural chemistry, petroleum industry and ways of better exploitation of the vast mineral resources of his country to improve the lot of the common Russians. Mendeleev was one of the earliest to realise the importance of petroleum as a raw material for petrochemical industries. He investigated into the origin of petroleum and was one of the first to come to the conclusion that it was biogenic in origin and was found under great depths on earth. *"The capital fact to note is that petroleum was born in the depths of the earth, and it is only there that we must seek its origin"* (From Mendeleev [10]). He was responsible for establishing the first oil refinery in Russia. Mendeleev felt strongly that '*burning petroleum as a fuel would be akin to firing up a kitchen stove with bank notes'* (From Moore, Stanitski & Jurs [11]).

In late nineteenth century, there was political turmoil in Russia which resulted in student unrest. Mendeleev resigned from the University of St. Petersburg in 1893 in protest against the resultant curbs on academic freedom and government's harsh treatment of students. Fortunately, friends at the Czar's court saved him. Instead of being sent to Siberia, he was appointed director of the Bureau of Weights and Measures. He was requested to fix the new state standard for the alcohol content of vodka. His recommendation that vodka should have 40% alcohol by volume was accepted by the government and was incorporated into Russian law in 1894. As director of weights and measures, Mendeleev introduced the metric system of measurement in Russia. Mendeleev was also instrumental in establishing the Russian Chemical Society in 1868/69.

Honors

Mendeleev was honoured by scientific bodies of many countries. Some of the honours he received are: Davy Medal of the Royal Society, London (1882) (with Lothar Meyer), Fellow of Royal Society of Edinburgh (1888), Copley Medal of the

(From "Dmitri Mendeleev", Wikipedia)

Royal Society, London (1905), elected as member of the Royal Society, London and of the Royal Swedish Academy of Sciences (1905), American Philosophical Society (1906). He was elected to the Academy of Arts of Russia for both his painting and insightful critique of contemporary painting. He received Honoris Causa degrees from universities around the world. Artificial element in the actinide series of atomic number 101 was named *mendelevium* in his honour. A large lunar impact crater located on the far side of the Moon is named after him.

(From "Dmitri Mendeleev", Wikipedia)

Sculpture in honor of Mendeleev and the periodic table, located in Bratislava, Slovakia

Mendeleev the man

Mendeleev was a Renaissance man with a larger than life, flamboyant personality. He was an icon both to students and to the oppressed peasants. He understood the needs of both as he himself rose above his background of poverty and fought all odds to become one of the intellectual giants of his time. Mendeleev was a passionate man who lived even his private life to the full on his terms. His first marriage to Feozva Leshcheva in 1863 was not happy and he spent more time doing science than with his young family. According to a story doing rounds at that time, annoyed Feozva asked Mendeleev if he was married to her or to science. In response, Mendeleev is reported to have said that he was married both to her and his science and if it was bigamy, then he was married to science! After 19 years of marriage he divorced Feozva in January 1882, to marry Anna Ivanova Popova, his niece's best friend. Even though Anna was very much younger than Mendeleev, she had great influence over him. When the church complained to the Czar about his second marriage, Mendeleev was so famous that the Czar seems to have said *"Mendeleev has two wives, yes, but I have only one Mendeleev"* (From Farber [12]). As years went by, Mendeleev cared less and less about his appearance. It was said that he would shave or trim his beard and cut his hair only once a year in spring. Neither his wife Anna nor the czar himself could persuade him to cut his hair!

Mendeleev was a man of principles and stood by them even at the cost of personal trouble. Mendeleev's resignation from the University of St. Petersburg is an example of his courage. He supported a student protest and their petition. When the minister of education not only refused to accept the petition but advised Mendeleev to confine himself only to teaching and not get involved with students and politics, he quit his post as a matter of principle on 17th August 1890.

From his youth, Mendeleev fought against injustices. He was also a great believer in freedom of spirit. He exhorted his fellow countrymen to guard their freedom at all cost. His exhortation to students in his last lecture at the University is the most fitting epitaph to this great Russian scientist and humanitarian. *"I have achieved an inner freedom. There is nothing in this world that I fear to say. No one nor anything can silence me. This is a good feeling. This is the feeling of a man. I want you to have this feeling too - it is my moral responsibility to help you achieve this inner freedom. I am an evolutionist of a peaceable type. Proceed in a logical and systematic manner"* (From Posin [2]). In his later years, Mendeleev used his prestige and influence to fearlessly speak out against injustices.

Mendeleev missed the Nobel Prize in 1906 in spite of the near unanimous recommendation of the Chemistry committee. Unfortunately, he died in 1907.

Mendeleev remained a greatly loved and admired Russian. He breathed his last on January 20, 1907, aged 73, *'while listening to a reading of Jules Verne's Journey to the North Pole' by his wife, Anna* (From Harrow [3]).

References

1. C. P. Snow, [Upon hearing the Periodic Table explained in a first-term university lecture] *"The Search"*, 1958, p25.
2. D. Q. Posin, *"Mendeleev, The Story of a Great Chemist"*, Whittlesey House, New York, 1948.
3. B. Harrow, *Eminent Chemists of Our Time, 2nd Ed., Van Nostrand, New York, 1927, pp. 18-40; 273-285".*
4. W. A. Tilden, "*Famous Chemists, The Men and their Work"*, George Routledge and Sons Ltd, London, 1921.
5. Dmitry Ivanovich Mendeleev, *'Principles of Chemistry', 1905, Vol. 2, 17-18.*
6. *(["The Correlation of the Properties and Atomic Weights of the Elements"],Zh. Russ. Khim. Obshch., 1869,1, No. 2/3, 60-77 (65)).* Masanori Kaji, *D. I. Mendeleev's concept of chemical elements and the principles of chemistry,* Tokyo Institute of Technology, Bull. Hist. Chem., Volume 27, Number 1 (2002), pg10.
7. Strathern, Paul. "*Mendeleyev's Dream: the Quest for the Elements*" Berkeley Books, 286-87, 2000.
8. D. Mendeleev, "The Relations of the Properties to the Atomic Weights of the Elements" Zeitschrift fur Chemie, 1869.
9. Eric John Holmyard, *"Makers of Chemistry'* Clarendon Press, 1929.
10. Dmitri Mendeleev, quoted from *L'Origine du pétrole. Revue Scientifique 2e Ser VIII, 1877.*
11. John W. Moore, Conrad L. Stanitski, Peter C. Jurs. *Chemistry: The Molecular Science*, Volume 1. Brooks/Cole, 2002.
12. E. Farber (Ed.), H. M. Leicester, Great Chemists, ("Dmitrii Ivanovich Mendeleev", "Encyclopedia of World Biography. 2004. Encyclopedia.com. 27 Jan. 2015).

Dmitri Mendeleev, source: Texas Public Library Archives, Author: Historical and Public Figures Collection (From Wikipedia, The free encyclopedia)

"Periodic table by Mendeleev, 1869" by NikNaks, Licensed under Public Domain via Wikimedia Commons - http://commons.wikimedia.org/wiki/File:Periodic_table_by_Mendeleev,_1869.svg

Mendeleev's Periodic Table, 1872, Source: Annalen der Chemie, Author: Dimitri Mendeleev. In public domain (From http://wiki.chemprime.chemeddl.org)

The sculpture of Dmitri Mendeleev in Saint Petersburg next to his Periodic Table on a wall of D.I.Mendeleyev Institute for Metrology, 26 June 2009. Author: Heidas. (From Wikipedia, The free encyclopedia)

"Mendeleyev gold Barry Kent" by Robert Wielgórski (Barry Kent) Licensed under CC BY-SA 3.0 via Wikimedia Commons - http://commons.wikimedia.org/wiki/File:Mendeleyev_gold_Barry_Kent.JPG

"Periodic table monument" by http://www.flickr.com/people/mmmdirt/ - http://www.flickr.com/photos/mmmdirt/279349599. Licensed under CC BY-SA 2.0 via Wikimedia Commons http://commons.wikimedia.org/wiki/File:Periodic_table_monument.jpg

9. JACOBUS HENRICUS van't HOFF (1852–1911)

First Nobel Laureate in Chemistry

"The work of van't Hoff is indissolubly woven in the texture of the chemistry of to-day. Whether we are organic chemists, inorganic chemists, or physical chemists, we constantly utilize and apply his ideas, reap the benefit of the intense thought he devoted to the fundamental problems of our science. This is his splendid and enduring memorial. Nothing can add to it, nothing detract from it" (From Walker [1]).

Early years

van't Hoff, Jr. was one of the seven children of the physician father Jacobus Henricus van't Hoff, Sr. and Alida Jacoba Kolff. Young van't Hoff grew up in a liberal household where interest not just in science but also in poetry and philosophy was encouraged. van't Hoff, Jr. shared with his father love for Byron's passionate poetry. During his youth, van't Hoff Jr. came under the influence of the French sociologist and philosopher, Comte. This had a long lasting influence on him and enabled him to consider everything including nature's mysteries from a philosophical point of view.

(From "J. H. van't Hoff", Wikipedia)

In an effort to prepare students to acquire skills required in a fast developing industrial economy, the Dutch government undertook reforms in education in an extensive manner during the decade of 1860. At the core of these reforms was setting up of high schools where importance was to be given to the study of science and mathematics to encourage students to choose a career in industry. van't Hoff was one of the earliest to get the benefit of this. van't Hoff joined the High School in Rotterdam where he got interested in doing chemical experiments. He gained access to the chemistry laboratory often after school hours and during school vacations by slipping through basement windows. When this was discovered and reported by the school authorities to his father, though initially angered by his son's misdemeanor, he realized that young van't Hoff should not be penalized for his interest in science. He gave space at home for his son to set up his own laboratory. Beginning in 1869, van't Hoff joined the Technical University in Delft to study chemistry and later joined the University of Leiden to do mathematics and physics. While still a graduate student, van't Hoff went to the University of Bonn (Germany) in 1872 for a year to study chemistry under the famous organic chemist August Kekule. There he came across Kekule's idea that the four bonds of carbon formed a tetrahedral structure (a possibility also proposed by Alexander Butlerov, a Russian chemist, Pasteur (optical isomerism - lactic acid), and Wislicenus). Kekule had proposed that a carbon atom was tetravalent as it combines with four other atoms to form an organic molecule. In the two dimensional model of Kekule, it was a linear compound with the carbon atoms situated at right angles to each other. van't Hoff got interested in this hypothesis and decided to investigate it further on his own. Encouraged by Kekule, he moved from the University of Bonn to École de Medicine in France in 1873 to study chemistry with Charles-Adolphe Wurtz where he met Le Bel. Interestingly, both of them were working independently on the same problem.

Tetrahedral Carbon

In September 1873, van't Hoff enrolled at Utrecht University and obtained his Ph.D when he was 22 years old. In 1874, he published a paper of 12 pages of text and one page containing only drawings titled *Voorstel tot Uitbreiding der Tegenwoordige in de Scheikunde gebruikte*

Structuurformules in de Ruimte, etc. (*Proposal for the Extension of Current Chemical Structural Formulas into Space, together with Related Observation on the Connection between Optically Active Power and the Chemical Constitution of Organic Compounds*), a rather long title for a short pamphlet in a Dutch journal. Using his drawings, van't Hoff was able to show theoretically that the four valences of the carbon atom were directed toward the four corners of a regular tetrahedron in space. Experimental proof came from others later. van't Hoff's paper was translated in 1875 into French with a short title as *"La chimie dans l'espace"* (Chemistry in Space). However, the scientific community became aware of his revolutionary theory only when its German version got published in 1877 under the title *"Die Lagerung der Atome im Raume"*. van't Hoff's hypothesis explained some cases of isomerism which had puzzled chemists and showed how the ability of certain chemical compounds to rotate a plane of polarized light (optical activity) was probably associated with the asymmetric nature of the carbon atom.

van't Hoff's work clearly showed that molecular structures of many organic molecules that were being studied by leading chemists at that time had indeed a three-dimensional shape and the three dimensionality was not just a means of conceptualizing molecules. This path breaking theory laid the foundation for stereochemistry or chemistry in space and brought van't Hoff into the centre of scientific limelight. Thanks to van't Hoff, chemistry became three dimensional! (Joseph Achille Le Bel of France, who was a coworker along with van't Hoff in Wurtz's laboratory, published independently the same theory, though in a slightly different form).

The ideas propounded by Le Bel and van't Hoff (aged 26 and 22 respectively) were different both in their starting points and approach. As van't Hoff himself noted *"On the whole, Le Bel's paper and mine are in accord; still, the conceptions are not quite the same. Historically the difference lies in this, that Le Bel's starting point was the researches of Pasteur, mine those of Kekule. My conception is ……a continuation of Kekule's law of quadrivalence of carbon with the added hypothesis that the four valencies are directed towards the corners of a tetrahedron at the centre of which is the carbon atom* (From van't Hoff [2])... *all the compounds of carbon which in solution rotate the plane of polarized light possess an asymmetric carbon atom"* (From van't Hoff [3]). van't Hoff approached the problem from structure theory.

While his revolutionary theory catapulted him to fame, many in the scientific community criticized van't Hoff for not providing experimental evidence to support his theory. His worst critic was Adolf Kolbe, editor of the *Journal für praktische Chemie*. He was vitriolic in his criticism. His intemperate response "*two virtually unknown chemists, one of them at a veterinary school and the other at an agricultural institute, pursue and attempt to answer the deepest problems of chemistry which probably will never be resolved (especially the question of the spatial arrangement of atoms), and moreover with an assurance and an impudence which literally astounds the true scientist*" (From Rocke [4]), was published in the *Journal für praktische Chemie*. In his personal attack on young van't Hoff who dared to question the beliefs held by senior chemists he said: "*A Dr. J. H. van't Hoff, of the Veterinary School of Utrecht, finds, it seems, no taste for exact chemical research. He has considered it more convenient to mount Pegasus (apparently on loan from the Veterinary School) and to proclaim in his "La chimie dans l'espace" how, during his bold flight to the top of the chemical Parnassus, the atoms appeared to him to be arranged in cosmic space*" (From Rocke [4]).

Ladenburg, another student of Kekule, criticised the idea of atoms in space. He said *"van't Hoff introduces into the formulas that I and most chemists purposely keep out of ..., namely, spatial representation... A formula must give account of composition, molecular weight and the way atoms are linked together"* (From Gadre [5]).

This violent criticism was also probably influenced by the fact the atomic theory was still considered a mere speculation. Ironically, the strong criticism of his theory, instead of destroying van't Hoff scientifically only made his theory more well known as even his critics had to first read it before criticizing it! van't Hoff had opened up a new area of stereochemistry which is important across many disciplines. van't Hoff published in 1894 a more detailed second edition of "The Arrangement of Atoms in Space" Wislicenus in the preface to this acknowledged the seminal importance of this paper in which he wrote *"it (the theory) has already affected to the full all that can be affected by any theory, for it has brought into organic connection with the fundamental theories of chemistry facts which were before incomprehensible and apparently isolated, and has enabled us to explain them from these theories in the simplest way……and has started in our science a movement full of significance, in a certain sense, indeed a new epoch"* (From van't Hoff [2]).

> van't Hoff had placed the logo of his favourite bar and student association on his dissertation. This had an unexpected fallout. Even though his paper had brought fame and the university realized he was a genius, van't Hoff was not appointed as a professor as the faculty feared that he would be partying instead of working. A secondary school would not hire him because he seemed absent minded, sloppily dressed and totally absorbed in his work. van't Hoff had to work as an university lecturer without any salary until he got a paid job at the veterinary school. Finally, in 1878 he became a professor of chemistry, mineralogy, and geology at the University of Amsterdam where he continued to work for 18 years before moving to the Academy of Sciences in Berlin for a named chair in 1896. The main reason for moving to Berlin was, he could not cope with his duties which included giving elementary lectures and examining a large number of students which left him too little time for research. He remained there till the end of his life.

van't Hoff's novel proposal that isomerism could be explained by the asymmetrical carbon atom had far reaching consequences. He showed that if four single covalent bonds of four different atoms or groups of atoms A, B, C, and D connect to a carbon atom, then two

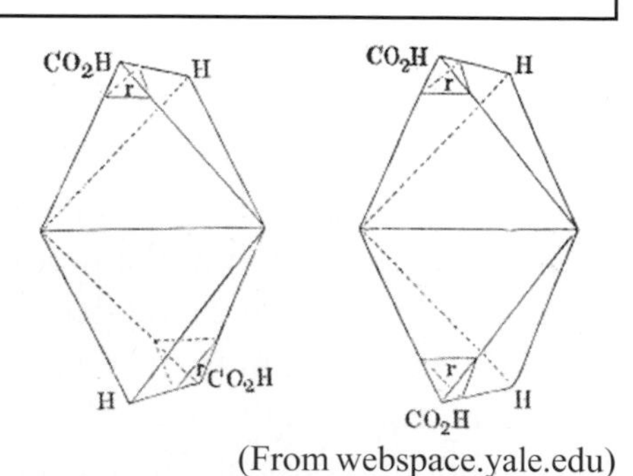

(From webspace.yale.edu)

different atomic arrangements are possible. This insight laid the foundation for stereochemistry and proved crucial to understand organic chemistry and to the study of three-dimensional nature of organic compounds.

van't Hoff devised a novel method to disseminate his idea of three-dimensional molecular properties to contemporary chemists. He sent them three-dimensional paper models of tetrahedral molecules!

Substituted ethane models made by van't Hoff for his fellow student G. J. W. Bremer. The set is preserved in the Leiden history of science museum.

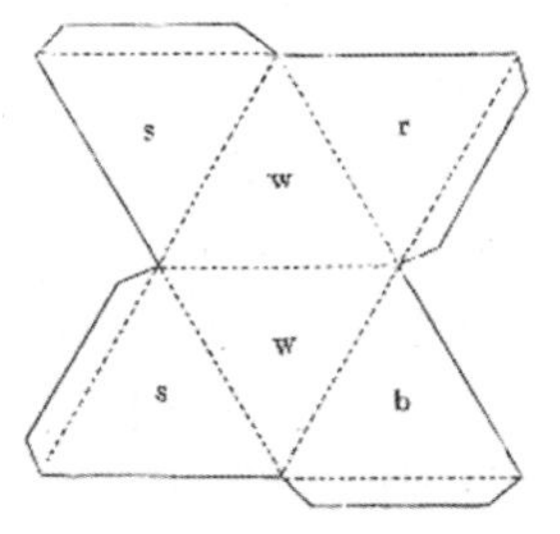

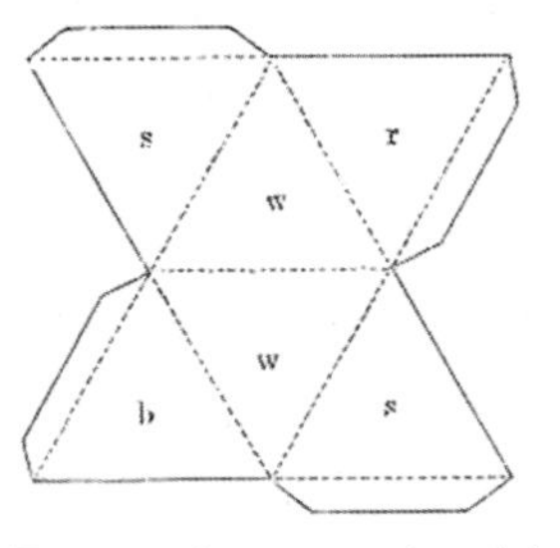

(From webspace.yale.edu)

(From www.dataphys.org)

In 1872, van't Hoff made a set of 3-D paper models of tetrahedral molecules

Foray into physical chemistry

Even though van't Hoff achieved recognition through organic chemistry, by late 1870s, he was no longer interested in studying organic molecular structures. His focus shifted to molecular transformations. He got deeply involved in studying why chemical reactions occur at widely different rates. In order to understand chemical equilibrium and chemical affinity, he began a decade-long research in thermodynamics and chemical kinetics. He developed a mathematical model to explain rates of chemical reactions based on the variation in the concentration of reactants with time. He derived an equation that gave the relation between the heat of reaction and equilibrium constant. This equation is widely known as the van't Hoff equation. In 1884, he published his famous book "Études de Dynamique Chimique" ("Studies in Chemical Dynamics") based on his research findings to explain chemical equilibrium and chemical affinity. In this book (which marked his foray into the emerging branch of physical chemistry), he put forth the principles of chemical kinetics for the first time and applied thermodynamics to explain chemical equilibria. He proposed a new method for determining the order of a chemical reaction. He also showed that when a system is subjected to temperature variations, heat was developed when the temperature was lowered and heat was absorbed when the temperature was increased. He thus explained the effect of temperature on equilibrium of reactions. Le Chatelier showed the applicability of this relationship and this is now known as van't Hoff- Le Chatelier Principle. van't Hoff deduced that when a reversible and isothermal chemical reaction occurred, the maximum external work done could be used as a measure of chemical affinity.

$$(- dc/dt) = kc^n$$

van't Hoff pointed out that chemical kinetics was different from chemical thermodynamics. Hermann von Helmholtz, German physicist had put forth a similar theory in 1882. He called such external maximum work obtained 'free energy'. van't Hoff is credited with introducing the idea that the driving force for chemical reactions was chemical affinity. Furthermore, he showed how concepts such as chemical equilibrium resulting when the rates of forward and reverse reactions are equal (the concept of dynamic equilibrium), the ratio of concentrations of starting materials to products at equilibrium (concept of the equilibrium constant), effect of the concentration of substances on the rates of reactions (concept of the law of mass action) together formed the basis for a coherent model for understanding the nature of chemical reactions. He was able to show how the rates of chemical reactions was affected by pressure, temperature and the concentration of the reactants, and how by applying a mathematical equation governing the final equilibrium state, it was possible to calculate the heat produced in a chemical reaction. By recognizing the relationship between equilibrium and the heat of reaction, van't Hoff was able to define chemical affinity.

van't Hoff defined chemical affinity in terms of the maximum external work done in a chemical reaction under constant temperature and pressure.

van't Hoff came to know from his colleague Hugo de Vries (the famous Dutch botanist at the University of Amsterdam) about the work of the botanist Pfeffer on osmosis in plant cells to understand nutrition process in plants. Pfeffer had discovered that to prevent the solvent passing through to the solution on the other side of the semi permeable membrane, excess pressure on the solution was required. This pressure was called osmotic pressure. Pfeffer had discovered the relation between the osmotic pressure of a solution and its concentration and temperature. van't Hoff was able to grasp the significance of this finding and its similarity to the ideal gas laws. He applied the principles of thermodynamics to show how osmotic pressure was directly proportional to the concentration of the solution. Extending this idea, he showed that the ideal gas law ($PV = RT$) applicable to gases was applicable to dilute solutions as well. He published a series of papers between 1886 and 1887 on this subject. He showed that the osmotic pressure in dilute solutions was proportional to the concentration (c) and the absolute temperature and that mathematically it was different from the formula for the ideal gas. Instead of $P = RT/V$, it is, $P = cRT$.

By combining chemical dynamics, thermodynamics with physical determination to understand the various aspects of complex chemical reaction, van't Hoff laid the foundation of physical chemistry.

van't Hoff remarked *"Whereas the chemico-chemists always find in industry a beautiful field of gold-laden soil, the physico-chemists stand somewhat farther off, especially those who seek only the greatest dilution, for in general there is little to make with watery solutions"* (From Johnson [6]).

One of the first physical chemists

"We can distinguish three groups of scientific men. In the first and very small group we have the men who discover fundamental

relations. Among these are van't Hoff, Arrhenius and Nernst" (From Bancroft [7]).

During the 1880 decade, the three chemists, Ostwald of Germany, Svante Arrhenius of Sweden and van't Hoff of Netherlands gave a new direction to physical chemistry based on thermodynamics. Till 1887, physical chemistry was not recognized as a separate branch of chemistry and there was no separate journal for the subject. To fill this void, van't Hoff and Ostwald started the journal "Zeitschrift für physikalische Chemie" (Journal of Physical Chemistry).

While at Berlin, van't Hoff turned his attention to the study of the salt deposits at Stassfurt, Germany. He wanted to find out under what conditions oceanic salt deposits were formed - another problem in chemical equilibrium. In order to understand the conditions behind the precipitation of salts, van't Hoff viewed the deposition process as an equilibrium between the solution and solid phases of the components in water at a constant temperature. This work was published in 1905 and 1909 as the two-volume *Zur Bildung der ozeanischen Salzablagerungen* ("On the Formation of Oceanic Salt Deposits").

Honours

van't Hoff was awarded the first Nobel Prize for Chemistry in 1901 for his discovery of the laws of chemical dynamics and osmotic pressure in solutions.

van't Hoff was recognized for his outstanding contributions to chemistry world over. Among the many honours and recognitions, he received are: Member of the Royal Netherlands Academy of Sciences (1885), Honorary Member of the Royal Dutch Academy of Sciences, (1892), Royal Academy of Sciences, Gottingen (1892), Davy Medal of the Royal Society, London (with J. A. Le Bel) (1893), Chevalier de la Legion d'Honneur, France (1894), Senator der Kaiser-Wilhelm-Gesellschaft (1911), Member of the French Academy of Sciences (1905), The Helmholtz Medal of the Prussian Academy of Sciences (1911).

He received Honoris Causa degrees from Harvard University and Yale University (1901), Victoria University, Manchester, UK (1903), Heidelberg University, Germany (1908).

The unique genius of van't Hoff

van't Hoff made fundamental contributions to unify important aspects of chemistry such as chemical kinetics, thermodynamics and emphasized the importance of physical measurements. van't Hoff's greatness lies in his unique ability to clearly see the crux of fundamental relationships and principles in the vast experimental data recorded by other scientists. A typical example of this gift was how he perceived instinctively the relationship between concentration and pressure and its similarity to the gas law from the data accumulated by Pfeffer in his studies on osmotic pressure in plant cells. He was able to formulate fundamental theories based on the observations made by others.

Most of van't Hoff's contributions were theoretical, rather than experimental. In an obituary Bancroft wrote in 1911:
"*In his whole life, he never made what would be called a very accurate measurement, and he never cared to. I remember his saying to me eighteen years ago, "How fortunate it is that there are people who will do that sort of work for us*!" (From Vaze [8]).

The end

van't Hoff contracted tuberculosis around 1906 and had to spend his last years in and out of a sanatorium. Neither the serious illness nor being confined to a sanatorium could dampen his zeal for research and writing papers. Even when he was being treated at the sanatorium, he persuaded the doctors to allow him to write whenever he was feeling well enough. As if to mock his illness, he wrote and published a paper in 1908 in the journal Biochemische Zeitschrift titled 'Sanatoriums Betractung' (View of the Sanatorium) in which he compared the relationship between heat and work while bedridden and while leading an active life. His last publication was in 1909 -1910 when he published his work on enzymes as catalysts in the formation as well as decomposition of glycosides. Many believe that this was the beginning of enzyme chemistry. He lost his battle against tuberculosis and died on March 1, 1911 at Stiglitz, Germany.

References

1. James Walker, van't Hoff Memorial Lecture, J. Chem. Soc., Trans., 1913,103, 1127-1143.
2. Jacobus Henricus van't Hoff, Edited and translated by Arnold Eiloart, *The Arrangement of Atoms in Space*, Cambridge University Press, 2014.
3. Van't Hoff on Tetrahedral carbon – Chemteam.info, http://www.chemteam.info/Chem-History/Van%27t-Hoff-1874.html) (From the article A suggestion looking to the extension into space of the structural formulas .... of organic compounds, Jacobus Henricus van 't Hoff, Archives neerlandaises des sciences exactes et naturelles, vol. 9, p. 445-454 September 1874).
4. Alan J. Rocke, *The Quiet Revolution: Hermann Kolbe and the Science of Organic Chemistry*, University of California Press, 1993.
5. Shridhar R. Gadre, 'Century of Nobel Prizes', *1901 Chemistry Award: Jacobus Henricus Van't Hoff (1852-1911)*, Resonance, p 36-43, December 2001.
6. Jeffrey Allan Johnson, *The Kaiser's Chemists: Science and Modernization in Imperial Germany,* University of North Carolina Press, 1990.
7. Wilder D. Bancroft, *Wilhelm Ostwald, the great protagonist*. Part II, J. Chem. Educ., vol.10, Issue 10. October 1933. pp. 583-646
8. Ravinder Vaze, *Jacobus Henricus Van't Hoff, The chemical Thinker*, Feature article, Science Reporter, April 2011.

Dr. J. H. van't Hoff, *The Arrangement of Atoms in Space (1874-77), (Source:https://webspace.yale.edu/chem125/125/history99/6Stereochemistry/vanthoff/tetrahedra.html).*

Dragicevic, Pierre and Jansen, Yvonne, *List of Physical Visualizations*, 2012, (Source:\href{http://www.dataphys.org/list}{www.dataphys.org/list}).

10. EMIL FISCHER (1852–1919)

A multi-faceted organic chemist

A great researcher, an innovative teacher, an institution builder and the first Nobel laureate in organic chemistry

"*Emil Fischer represents a symbol of Germany's greatness*" (From Harries [1(a)]).

Early years

(From "Hermann Emil Fischer",Wikipedia)

Hermann Emil Fischer, son of Laurenz Fischer and Julie Poensgen Fischer, was born on October 9, 1852 in Euskirchen (Cologne district), Germany. His father had a successful dye making and lumber business. In addition, he also ran a profitable brewery. Being the youngest and only son with five elder sisters, Emil was pampered and had a happy childhood. His early education included three years of private tutoring and two years at the local Higher Public School. This was followed by two years at the gymnasium at Wetzlar and two more years at Bonn. He was good at his studies and passed out of the gymnasium at Bonn with distinction in 1869. He was forced to join his father's business much against his will. His task was to assist his father to solve some chemistry related problems in dye making and brewing. After his none too impressive performance in the successful family business, his father apparently decided that his son was too stupid to succeed in business and that he would do better as a student (From www.nobelprize.org [2]). Released from the drudgery of working in the family business, Fischer joined the University of Bonn to study chemistry where he attended lectures of some of the leading German chemists such as Kekulé, Engelbach and Zincke. In addition to chemistry, he also attended physics lectures of August Kundt and mineralogy lectures of Paul Groth. The turning point in Fischer's academic destiny was the result of the unexpected move to the newly established University of Strasbourg in 1872. Emil Fischer who now desired to study physics was encouraged by his cousin Otto Fischer to move with him to the University of Strasbourg to work under Professor Rose. This move changed Fischer's professional career. Instead of working under Professor Rose there, Fischer came under the influence of the famous chemist of that time Adolf von Baeyer, the newly appointed director of the chemical institute. Baeyer convinced Fischer to take up chemistry as his life's calling. Thus began Fischer's tryst with chemistry. He took up the study of phthalein dyes for his Ph.D. thesis. Working under Baeyer, Fischer obtained his doctorate in 1874 for his thesis on fluorescein and orcin-phthalein from the University of Strasbourg. After obtaining his Ph.D., Fischer continued research at Strasbourg. While working with von Baeyer, he discovered a compound which played a crucial role in his future research on sugars - phenylhydrazine. He also determined its structure.

On the flip side of the discovery of phenylhydrazine, it was perhaps at the root of the serious gastritis problem that he developed as a result of inhaling it continuously while working in a unventilated laboratory during his research on sugars.

Fischer moved to the University of Munich as Professor Baeyer's assistant in 1875 where he continued his research on hydrazines. He was appointed professor of chemistry at the University of Erlangen in 1881 and he later moved to the University of Wurzburg in 1888. He spent his happiest years there during 1888-92. Finally, he moved as the chair of chemistry at the University of Berlin in 1892, after the retirement of Professor A. W. Hofmann and remained there till his death in 1919.

Purines and sugars

The sixteen years between 1882 to 1898 were the most productive years of Fischer's research. This period witnessed his monumental work on purines and sugars. His Nobel citation says *"Nobel Prize was awarded to Fischer in recognition of the extraordinary services he has rendered by his work on sugar and purine syntheses"* (From www.nobelprize.org [3]).

Fischer discovered the answer to the question "What is common to tea, coffee and chocolate?" Fischer began his investigations in 1881-82 at the University of Erlangen on the active chemicals present in coffee, tea and cocoa and found that they all had caffeine and theobromine. On further investigation, he discovered in 1884 that substances of vegetable origin (caffeine), of animal origin (in the animal excretion such as uric acid) and compounds like adenine and xanthine all shared a parent system of a nitrogen containing base with a bicyclic structure. He also found that it was possible to derive these compounds from one another. He termed this substance (which he at first thought was just hypothetical) purine. It is one of the important components of nucleic acids (DNA). He succeeded in synthesizing it in 1898. This discovery opened a new branch in organic chemistry - purine chemistry. This research also had a utilitarian value as it was responsible for significant advances in the drug industry of Germany. Fischer also identified and synthesized many amino acids.

When Fischer turned his attention to the study of sugars in 1884, little was known about the chemistry of naturally occurring sugars except that the four variants of simple sugars – glucose, galactose, fructose and sorbose-shared the same molecular formula $C_6H_{12}O_6$ and certain other common features. Fischer decided to determine the structure and configuration of sugars. He started his work with glucose as this was the simplest. To determine its chemical properties such as its solubility and optical activity, Fischer had to first determine its carbon skeleton and the nature and the spatial location of its functional groups. As not much was known about what reagent would be ideal, Fischer decided to use phenylhydrazine which he had synthesised in the early 1870s at Strasbourg. This was a masterstroke. His path breaking discoveries in sugars was possible mainly because of the vital role of phenylhydrazine, as sugars which had proved difficult to purify and characterize till then, reacted with phenylhydrazine to form highly crystalline osazones which could be purified easily. Osazones also formed distinctly coloured crystals with different shapes (for example needle shaped crystals with glucose and sunflower shaped ones with maltose) which helped in identifying the sugars. He repeated the experiment with fructose and mannose and found that they also yielded the same osazone. In a leap of insight, Fischer realised that by applying van't Hoff's concept of the tetrahedral geometry of the carbon atom resulting in stereoisomers, it was possible to find that the sugars were actually spatial isomers. He proved that the four carbon atoms in a glucose molecule could have 16 possible stereoisomers. He was successful in determining the structures of many sugars. He verified the structures by synthesizing each of the sugars in his laboratory. Between 1882 and 1896 Fischer and his coworkers produced in the laboratory a number of artificial analogues of naturally occurring sugars. He was the first chemist to synthesize glucose, fructose and mannose starting from glycerol in the laboratory and showed the chemical relation between glucose, fructose and mannose.

"Once a molecule is asymmetric, its extension proceeds also in an asymmetrical sense. This concept completely eliminates the difference between natural and artificial synthesis. The advance of science has removed the last chemical hiding place for the once so highly esteemed vis vitalis" (From Wolfrom [4]).

What did Fischer's beard have to do with his success in crystallizing sugars?

It is often joked among scientists that it was not phenylhydrazine but his luxuriant beard that played the crucial role in crystallizing sugars. It is said that as Fischer worked on his experiments, fine particles of chemicals would get embedded in his beard. When he shook his head when perplexed by the problem or in joy while watching the reaction in a glass dish containing the solutions, some of the chemical particles that fell into the solution would act as the nuclei for crystallization to start!

Fischer Projections

Fischer's research to understand the structure of sugars led him to develop in 1891 a novel way of representing the three dimensional structures of organic molecules as two dimensional drawings without changing either their structural authenticity or properties.

In the projection of a three dimensional carbon compound as a two dimensional drawing, Fischer used a pair of intersecting horizontal and vertical lines where the horizontal lines represented the atoms facing the viewer and the vertical lines atoms pointing away from the viewer. The horizontal lines are solid lines and the vertical lines are dashed lines. The carbon atoms are at the intersection of these two lines.

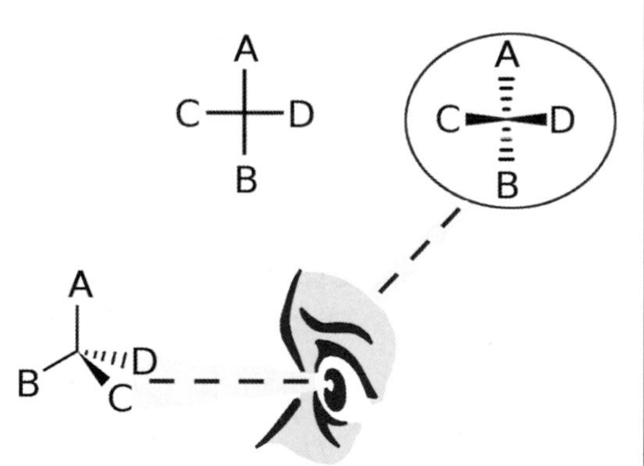

This method of structural projection conveys valuable stereochemical information to the reader. However, Fischer projections can be used only to represent molecules with stereogenic centers. In addition to being used extensively in organic chemistry and biochemistry to represent monosaccharide sugars and amino acids, it can also be used to differentiate optical isomers.

Enzyme action and the lock and key model

Fischer's work on sugars led him to study the relation between sugars and fermentation. One of the off-shoots of this work was his study on how the ferments or enzymes which caused the fermentation of sugars worked. While working on hydrolysis of glucosides (the isomers α–methylglucoside and β- methylglucoside), he discovered that they showed a distinct specificity for the enzyme reacting with them. Fischer recognised the reason for this specificity and proposed the "Lock and Key Model" to visualize the interaction between an enzyme (the lock) and the substrate (key). This model states that just like a lock and its key, an enzyme and the substance upon which it acts or its substrate must fit together precisely for a biochemical reaction to get unlocked or begin. Lock and key analogy has helped scientists to visualize the specificity of enzyme action and to get a mental picture of molecular recognition processes. It also contributed significantly to the development of organic chemistry, biology and drug development.

"Their (enzymes) specific effect on the glucosides might thus be explained by assuming that the intimate contact between the molecules necessary for the release of the chemical reaction is possible only with similar geometrical configurations. To give an illustration I will say that enzyme and glucoside must fit together like lock and key in order to be able to exercise a chemical action on each other. This concept has undoubtedly gained in probability and value for stereochemical research, after the phenomenon itself was transferred from the biological to the purely chemical field"- Emil Fischer (From Holmstedt & Liljestrand [5]).

Protein research

In 1899, Fischer decided to shift his focus to the study of proteins in order to understand their structural properties. Little was known about this important chemical substance except that it was made up of amino acids and only 13 amino acids were known. He developed more effective analytical methods for identifying and separating individual amino acids and discovered the cyclic amino acids namely proline and hydroxyproline.

Fischer was one the earliest chemists to understand the complex nature of proteins. He succeeded in establishing that a common type of bond held the two successive amino acids and named it peptide bond. Based on this knowledge, he synthesized dipeptides, tripeptides and polypeptides. He also prepared naturally occurring as well as synthetic proteins in his laboratory (e.g. Octodecapeptide), which had many characteristics of natural proteins. Though he had made hundreds of proteins in his laboratory by 1916, Fischer knew that he was only scratching the surface as millions of proteins were yet undiscovered.

Another major contribution of Fischer to protein chemistry was the introduction of fractional distillation at low pressure and temperature to purify esters of amino acids (important for peptide synthesis) and thus limit the decomposition of these esters when they are purified at high temperatures and normal pressure.

Fischer was involved in the discovery of barbiturates, used in the preparation of sedative drugs to treat insomnia, epilepsy, anxiety, and anesthesia. Barbital, the first sedative drug was launched in 1904 with the help of Fischer and Josef von Mering, a physician.

Honours

Emil Fischer received a number of honours for his fundamental contributions to organic chemistry. Some of them are: Davy Medal of the Royal Society, London, 1890, Nobel Prize in Chemistry (1902). He was the first Nobel Prize winner in organic chemistry. Prussian Order of Merit and the Maximilian Order for Arts and Sciences (1913) Helmholtz Medal for his work on protein chemistry and sugars (1909).

He received honorary doctorates from the Universities of Christiania, Cambridge (England), Manchester and Brussels.

His contributions to organic chemistry are so fundamental to the subject that many concepts and reactions are named after him. Some of them are: Fischer projection, Fischer oxazole synthesis, Fischer peptide synthesis, Fischer phenylhydrazine and oxazone reaction, Fischer reduction, Fischer-Speier esterification.

Fischer the man

Fischer was a quintessential experimental chemist. He strongly believed that any hypothesis had to be supported by experimental proof before it could be accepted and insisted on his students providing experimental proof of their work.

"You are urgently warned against allowing yourself to be influenced in any way by theories or by any other preconceived notions in the observation of phenomena, the performance of analyses and other determinations" (From Bergmann [1(b)]).

Apart from his contributions to chemistry, Fischer made enormous contributions to the growth of science in Germany. He was a great institution builder. He campaigned tirelessly for establishing scientific institutions. He was instrumental in establishing the Kaiser Wilhelm Society for the Advancement of Sciences in 1911 and the Kaiser Wilhelm Institute for Chemistry (which were renamed Max Planck Institutes). He was affectionately called the "secret prince of chemists" (From Nagendrappa [6]) for his unequalled contribution to higher education, unstinting support to scientific research and promotion of industry in Germany. Like many of his contemporary scientists, Fischer was also drafted by the military establishment during World War I. He was in charge of chemical industries to produce essential chemicals needed for the conduct of the war.

Fischer was a brilliant teacher even though he was not a gifted communicator. He had phenomenal memory and used this gift to memorise his entire lecture and was able to give what appeared like extempore lectures. He was a kind and empathetic person. He was admired and loved for his love of truth and his intuitive understanding of problems (both scientific and personal problems of others). He loved nature. While he was at the University of Würzburg, he enjoyed walking in the hills and the black forests.

Final years

As the World War I drew to a close, the futility of war left Fischer deeply disillusioned. The war was also a personal catastrophe as two of his sons died in the war. This tragedy and the deterioration of his health (with suspected stomach cancer) resulted in Fischer suffering extreme depression. He died on July 15, 1919 and was mourned by his students and admirers alike.

"Modern warfare is in every respect so horrifying, that scientific people will only regret that it draws its means from the progress of ..sciences. I hope that the present war [World War I] will teach the peoples of Europe a lasting lesson and bring the friends of peace into power" - Fischer (From Nagendrappa [6])

German Chemical Society instituted The Emil Fischer Memorial Medal in recognition of his services to chemistry.

References

1. Joseph S. Fruton, *Contrasts in Scientific Style: Research Groups in the Chemical and Biochemical Sciences*, Volume 191 of Memoirs Series, American Philosophical Society, p 167, 1990. (a) (Harries (1919a, p 843), (b) (Bergmann (1930), p.415.)
2. "Emil Fischer - Biographical". *Nobelprize.org.* Nobel Media AB 2014. Web. 6 Apr 2015. http://www.nobelprize.org/nobel_prizes/chemistry/laureates/1902/fischer-bio.html
3. The Nobel Prize in Chemistry 1902, *Nobelprize.org.* Nobel Media AB 2014. Web. 6 Apr 2015. http://www.nobelprize.org/nobel_prizes/chemistry/laureates/1902/
4. Melville L. Wolfrom, *Advances in Carbohydrate Chemistry*, Volume 21, Academic Press Inc. 1966.
5. B. Holmstedt, G. Liljestrand (eds.) *Readings in Pharmacology* (1963), The Macmillan Company, New York, p251.
6. G. Nagendrappa, *Hermann Emil Fischer: Life and Achievements*, Resonance, July 2011, p616 (606 -618).

Hermann Emil Fischer, 1895, Author- Atelier Victoria (Inh. Paul Gericke, gegr. 1894), (Wikimedia Commons)

11. WILHELM OSTWALD (1853–1932)

High priest of physical chemistry

"Ostwald was a great protagonist and an inspiring teacher. He had the gift of saying the right thing in the right way. When we consider the development of chemistry as a whole, Ostwald's name like Abou ben Adhem's leads all the rest ....." (From Bancroft [1]).

Early years

Wilhelm Ostwald was born on September 2, 1853, in Riga, Latvia, to Gottfried Wilhelm Ostwald, a master cooper and Elisabeth Leuckel. He had his early education in Riga. Wilhelm Ostwald inherited his love for music and humanities and skill in painting from his father. His father wanted Wilhelm to become an engineer but he chose to study chemistry when he joined the Dorpat University in 1872. Initially he was interested in studying liberal arts (music, philosophy etc) and in having a good student life. Ostwald seems to have realised the seriousness of his situation in the nick of time and got through the theoretical chemistry examination. He was admitted to Professor Carl Schmidt's laboratory to study chemistry and received the master's degree in 1876 and Ph.D in 1878 from Dorpat University. Subsequently,

(From "Wilhelm Ostwald", Wikipedia)

he got a job as part-time laboratory assistant which enabled him to continue his research there till he moved to Riga Polytechnic in 1881 as professor of chemistry. Ostwald soon established himself as an outstanding lecturer and undertook two important projects which propelled him into gaining public recognition. He published a book, "Lehrbuch der allgemeinen Chemie" (Textbook of general chemistry), and started the journal, 'Zeitschrift für Physikalische Chemie'. Both these efforts were instrumental in establishing physical chemistry as an independent branch of chemistry. Using apparatus built out of easily available and inexpensive materials, Ostwald did much of his important research far away from the major and well established centres of Europe. He remained at the Riga Polytechnic till he moved to Leipzig in 1887 to take up a special chair.

Even as a graduate student, Ostwald realized the need to change the way chemistry was taught in the universities and practiced by contemporary chemists. He is said to have remarked at his Master's exam "*Modern Chemistry is in need of reform*" (From Servos [2]). Ostwald noticed that while physics had already become a quantitative and rational discipline, concerning itself with finding answers to how and why of natural phenomena, chemistry in comparison was still in its infancy. He was impatient to modernize the nature of chemistry. In the process, Ostwald pioneered the approach of adapting physical measurements to solve issues of chemical dynamics and thus was primarily responsible for establishing the new school of physical chemistry.

Pioneer in physical chemistry

When Ostwald started his research, chemistry (spurred by Wohler's synthesis of urea in the laboratory) in Europe, and particularly in Germany, was mostly focused on investigating what new products could be synthesized from known chemical compounds. There was no understanding of the quantitative aspects of chemical reactions. Ostwald decided to tackle this through physical measurements and mathematical reasoning. For example, he devised a simple method to study the equilibrium in aqueous solution. When he heard that the Dutch chemist Thomson had used the thermochemistry route to study solutions, Ostwald realized immediately the significance of this approach and used it in his work. He stated sometime later, "*like lightning, the thought occurred to me that instead of heat development, every other measured quality of the solution could serve, provided the quality is sufficiently influenced by a chemical change*" (From Färber [3]). Subsequent to this insight, Ostwald measured changes in physical properties like volume, refractive index, and electrical conductivity that occurred during chemical reactions. He used the law of mass action proposed by Cato Guldberg and Peter Waagey of Norway to analyse the data to estimate the relative affinities of substances to explain and predict the way solutions behave when they are in dynamic equilibrium.

Ostwald held the view that his fortuitous choice of physical chemistry – which did not exist as a separate branch of chemistry at that time - was due more to his isolation from the mainstream research conducted in other parts of Europe than by any preference to the then unfashionable subject.

Ostwald - Arrhenius

In 1884, Ostwald received from Svante Arrhenius his doctoral thesis on the electrical conductivity of solutions in which he the explained the method he had used to determine the conductivity of dilute acids. He boldly stated that when salts, acids and bases are dissolved in water, they dissociate into their charged ions. Ostwald realised at once the importance of this theory. He surmised that if all acids contained the same active ion, then the differences in the concentration of active ions in each of the acids would cause differences in their chemical activities and also the differing degrees of dissociation of the acids would result in the differences in the concentration of active ions. Ostwald recognized that if the law of mass action were to be applied to the dissociation reaction, it would be possible to derive a simple mathematical relation between the degree of dissociation (α), the concentration of the acid (c), and an equilibrium constant specific for each acid (k) with the formula $\alpha^2/(1-\alpha)\,c = k$. To test his hypothesis, he measured the electrical conductivities of over 200 organic acids. The data he obtained from these measurements validated the theory of dissociation. This came to be known as Ostwald's Law of Dilution. Ostwald is also credited with introducing the concept of a mole. Around this time van't Hoff of Holland joined Ostwald and Arrhenius. van't Hoff added a new dimension to this emerging field – the theory of osmotic pressure.

> The triumvirate of physical chemistry- Ostwald, van't Hoff and Arrhenius - were called the "Ionists" for their seminal individual contributions to the theory of dissociation of electrolytes in liquids. The contributions of these three giants coalesced into the comprehensive theory of solutions grounded in thermodynamics and theory of dissociation and the new school of physical chemistry was born.

Father of physical chemistry

"Ostwald was absolutely the right man in the right place. He was loved and followed by more people than any chemist of our time" (From Bancroft [1]).

"We can distinguish three groups of scientific men. In the first and very small group we have the men who discover fundamental relations. Among these are van't Hoff, Arrhenius and Nernst. In the second group we have the men who do not make the great discovery but who see the importance and bearing of it, and who preach the gospel to the heathen. Ostwald stands absolutely at the head of this group. The last group contains the rest of us, the men who have to have things explained to us" (From Bancroft [1]).

Even though Ostwald (Law of dilution), van't Hoff (theory of osmotic pressure) and Arrhenius (theories of electrolytic dissociation and of free ions) had all made original contributions to the subject, Ostwald is considered to be the father of physical chemistry for collating individual pieces of the jigsaw puzzle to produce the complete picture of why and how the dissociation of electrolytes occurred. Nernst, Ostwald, van't Hoff and Arrhenius were the intellectual giants of the glorious era when physical chemistry emerged as a discipline. Each of them made path-breaking contributions to this nascent field. Ostwald had the gift of expression and used it to explain the uniqueness of the new discipline to solve problems in other branches of chemistry with such lucidity that contemporary professional chemists as well as students understood the subject. A large number of chemists from different parts of the world came to study under him. Ostwald is also credited with systemizing this new branch of chemistry. He wrote about the salient features of this new discipline in text books on other branches of chemistry such as general, inorganic and analytical chemistry. He was able to revolutionize analytical chemistry by employing the theory of solutions and his own theory of indicators. He was also responsible for establishing the Leipzig Institute of Physical Chemistry, a world class centre for physical chemistry which soon became famous for training students in new experimental skills and ideas. Ostwald conducted research on various aspects of physical chemistry such as electrical conductivity, dissociation of acids and bases, mass action, rates of reaction and catalysis. Concept of chemical affinity was central to all these areas.

Catalysis

Catalysis had been observed in many chemical reactions long before Ostwald. As the concept of the rate of reaction was unknown at that time, Berzelius could not explain the phenomenon and had ascribed it to catalytic force. Ostwald did not agree that the phenomenon of catalysis was the result of catalytic force.

"The development of a rational view of the nature of catalysis was thus absolutely dependent on the creation of the concept of the rate of chemical reaction" (From Ostwald [4]).

Ostwald's work on catalysis had its origin in his attempt to link rates of reactions to the chemical affinity of the substances taking part in the reaction. During the same period, Ostwald and his students conducted many experiments to measure the rates at which esters hydrolysed when they reacted with mineral acids. Ostwald discovered that the concentration of the acid affected the rate at which the reaction proceeded towards the equilibrium state. As Ostwald had studied acid-base chemistry, he found that acids with high chemical affinity increased the rate of the chemical reaction. He soon realised that a catalytic reaction was caused by the addition of a foreign compound to the substances taking part in the reaction and this substance or the catalyst (from the Greek word katalusis, meaning dissolution – dissolving all obstacles to the chemical reaction taking place) was responsible for the change in the velocity of the reaction. Using the principles of thermodynamics, he showed that catalysts did not have any effect on the equilibrium constant as the rate constants of both the forward and backward reactions were altered to the same extent. He defined a catalyst as *"a substance which alters the velocity of a chemical reaction without appearing in the final products"* (From Partington [5] & Moore [6]) After his studies of the homogeneous catalysis, Ostwald started his investigations on heterogeneous catalysis.

Ostwald's greatest contribution to the area of catalysis was the invention of the process for the preparation of nitric acid from oxidation of ammonia in bulk. This process came to be known as Ostwald process. This discovery had far reaching implications both for chemical theory and fertilizer industry (where it is used even today).

Ostwald was proud of his work on catalysis.

In 1909 Ostwald was awarded the Nobel Prize in chemistry for his work on catalysis. In announcing the award, the presenter declared: *"Catalysis, which formerly appeared to be a hidden secret, has thus become ... accessible to exact scientific study"* - President of the Royal Academy of Sciences (From *Nobel Lectures in Chemistry* [4]).

As part of his Nobel lecture, Ostwald said *"It has pleased no less than surprised me that of the many studies whereby I have sought to extend the field of general chemistry, the highest scientific distinction that there is today has been awarded for those on catalysis. ....in my innermost being, I used to, and still do, consider this part of my work the one in which the personal quality of my method of work is most definitely shown up and which I therefore have more at heart than all the others"* (From Ostwald [4]).

Ostwald and energetics

"The special importance of thermodynamics in the development of what we now call classical physical chemistry……………led Ostwald finally to the conviction that energy was a fundamental reality, the underlying component of the physico-chemical world" (From Donnan [7]).

Ostwald's work on catalysis got him interested in measuring the energy changes that occur when chemical reactions take place i.e. the energy absorbed or released during chemical reactions. He was convinced of the supremacy of energy over everything in the world and named this approach as "energetics". He tried to understand all phenomena in terms of either transformation or transfer of energy. He firmly believed that matter was only *"a mirage which the mind creates to comprehend the workings of energy"* (From Sutton [8]) and *"What we call matter is only a complex of energies found together in the same place"* (From Encyclopedia.com [9]). He believed that science made a positive contribution to society and therefore any obstacle placed deliberately or otherwise in its path was a waste of social energy. His exhortation *"Do not waste energy, but transform it into a more useful form"* (From Holt [10]) became his rallying cry to society.

Rejection of atomic hypothesis

By the 1890s Ostwald became immersed in finding solutions to the theoretical and practical problems posed by the concept of energy. Around this time, the atomic hypothesis was widely accepted by scientists world over. Ostwald refused to accept the concept of atoms and molecules and called them only hypothetical particles. He proclaimed that energetics was best suited for expressing basic scientific truths. Perhaps because of this conviction, Ostwald did not accept atomic theory or even the concept of atoms till 1909 when irrefutable experimental evidence such as the measurement of the Avogadro number and the Brownian motion theory of Einstein forced him to finally accept the atomic hypothesis.

Ostwald Ripening

Ostwald's work on the nucleation of crystals in solutions resulted in what came to be known as 'Ostwald ripening'. During the course of his research, on solutions, he noticed that in the beginning dissolved particles in a solution formed small particles and with time the small particles further redeposit themsclves to form bigger particles. Ostwald predicted that as this process goes on, ultimately the solution will have only big particles; i.e. the small particles form the nutrients for the growth of large particles. This is now popularly known as 'Ostwald ripening'. Ostwald ripening is found in many physical processes.

Colours

Ostwald was interested in painting from his childhood and as he grew older, he mixed his own pigments to get the colours he wanted. This interest in painting led him to the realisation that some colours combined to give more appealing and harmonious colours than other combinations. He decided to make a systematic study of colour combinations. He aim was to replace the subjective and qualitative perception of colours with an objective and quantitative classification. He published a colour atlas called *Die Farbenfibel* (The Colour Primer) which contained 2500 different colours. He published the books, *Die Farbenlehre* (Colour theory), *Die Harmonie der Farben* (Harmony of the colours) and a periodical *Die Farbe* (Colour) on the subject of colours.

Views on Science

Ostwald was truly a man of science and believed in the universality of science. In a speech that he gave at the banquet for Past Presidents of the Chemical Society, he said "*Science is one land, having the ability to accommodate even more people, as more residents gather in it; it is a treasure that is the greater the more it is shared. Because of that, each of us can do his work in his own way, and the common ground does not mean conformity*" (From Crookes [11]).

He also understood the interdisciplinary nature of the subject and this understanding is reflected in his multi-pronged approach to find solution to specific scientific problems and this approach enabled him to make seminal contributions to a variety of fields. Ostwald was an atheist and he believed that science would soon attain the status of a religion - eternal, omniscient and perfectly good.

Ostwald was a great advocate of international scientific cooperation. He was instrumental in establishing Association of Chemical Societies, an international platform where scientists from different countries could meet and exchange ideas. He was also a prolific writer. In addition to more than five hundred research papers, he published 45 books. Among the textbooks he published, *Lehrbuch der Allgemeinen Chemie*, 1884 (Textbook of general chemistry), *Grundriss der Allgemeinen Chemie,* 1889 (Outlines of general chemistry) and *Hand- und Hilfsbuch zur Ausführung physikalisch-chemischer Messungen* 1893 (Handbook and manual of physico-chemical measurements) are noteworthy. He was also editor of many scientific as well as philosophical journals.

Honours

In recognition of his outstanding contributions to chemistry, several universities in Germany, UK and the USA conferred honoris causa degrees on him. He was also an honorary member of scientific societies of Norway, Sweden, The Netherlands, Great Britain, the USA and Russia. In 1899 he was made a "Geheimrat" (the highest title occasionally given to academics) by the King of Saxony. He was awarded the Nobel Prize in 1909 for chemistry.

Ostwald the man

"*Picture to yourselves a friendly enthusiastic man, with penetrating eyes, fresh colour, and reddish hair, moustache and beard, going the round of the research laboratories every day. If you had a difficulty, Ostwald had a solution to offer. If you had no difficulties, you probably got some new ideas. If you had any views on music, painting or philosophy, the Master was full of attention and would discuss them with you*" (From Donnan [7]).

Apart from being a pioneer in chemistry, Ostwald was a great human being. He combined outstanding scientific acumen with his love for students. These qualities made him an extraordinary mentor. *"He was loved and followed by more people than any chemist of our time"* (From Bancroft [1]). His generosity to young and struggling Arrhenius is a testimony to his extraordinary human qualities. When Ostwald learnt that Arrhenius had been awarded a consolation doctorate degree which made it impossible for him to get either a teaching or research position, he offered Arrhenius to join his group and thereby rescued him from being consigned to obscurity. That Arrhenius went on to become one of the greatest chemists of all time is entirely due to the magnanimity of Ostwald. Most of the early physical chemists of the U S were his students.

Ostwald was a workaholic who believed that *"Happiness is equal to work minus resistance"* (From Chymia [12]). He was a man of principles and when there was some discord at his university, he resigned from his position and retired to his country home "Energie" where he spent the rest of his life in the world of colours.

Ostwald was an idealist who believed in a world without war and border. As part of his efforts to build one world and one humanity, he even supported the movement to create a universal language, Ido which would be the glue that would hold different people together but to his great disappointment this had very few takers. He tried to build an organization appropriately called Die Brücke (The Bridge) to build bridges across rigid national boundaries as well as the rich cultural and intellectual achievements of different countries but unfortunately this dream also had no takers.

The end

After a short period of suffering from bladder and prostate troubles, Ostwald breathed his last peacefully in a Leipzig hospital on April 4, 1932 and was buried in the grounds of his private estate.

References

1. Wilder D. Bancroft, *Wilhelm Ostwald, the great protagonist*. Part II, J. Chem. Educ., vol.10, Issue 10. October 1933. pp583-646, (quoted on 612).
2. John W. Servos, *Physical Chemistry from Ostwald to Pauling: The Making of a Science in America*, Princeton University Press, 1990, p3.
3. Eduard Färber, *A study in scientific genius: Wilhelm Ostwald's hundredth anniversary,* J. Chem. Educ., 1953, 30 (12), p 600.
4. (Wilhelm Ostwald, *On catalysis)*, Nobel Lecture, *Nobel Lectures in Chemistry, 1901–1921*, World Scientific Publishing Co. Pte. Ltd., 1999.
5. As quoted in J. R. Partington, *A History of Chemistry*, Vol. 4 (1901), 599-600, ('Über Katalyse', *Zeitschrift für Physikalische Chemie* (1901), **7**, 995-1004)).
6. W. J. Moore, *Physical Chemistry* (5th Edition), Orient Longman Pvt. Ltd., India, 2004.
7. F. G. Donnan, *Ostwald memorial lecture*, J. Chem. Soc., 1933, 316 -332.
8. Michael Sutton, *The Father of Physical Chemistry*, 'Chemistry World', May 2003, Royal Society of Chemistry.
9. "Ostwald, Friedrich Wilhelm." Complete Dictionary of Scientific Biography. 2008. *Encyclopedia.com.* 25 Mar. 2015 <http://www.encyclopedia.com>
10. Ostwald, 'Der energetische Imperativ', in Annalen der *Naturphilosophie*, x (1911), 11317. (From Holt, Niles R. 'Wilhelm Ostwald's "The Bridge"', British Journal for the History of Science, vol. 10, Part 2, no. 35, July 1977, pp. 146-150).
11. Sir William Crookes, *The Chemical News and Journal of Industrial Science...,* Volume 78, 1898. (Chemical Society – Banquet to Past Presidents).
12. Chymia: Annual Studies in the History of Chemistry, Volume 2, University of Pennsylvania Press, 59, 1949.

12. SVANTE ARRHENIUS (1859–1927)

Some compounds dissociate, and some cause climate change

"Chemistry works with an enormous number of substances, but cares only for some few of their properties; it is an extensive science. Physics on the other hand works with rather few substances, ...but analyses the experimental results very thoroughly; it is an intensive science" (From Arrhenius [1]).

Svante Arrhenius changed the way chemists looked at aqueous solutions of acids, bases and salts by startling the scientific world with his theory of electrolytic dissociation. According to this theory, some chemical compounds when dissolved in water, dissociate into positively charged and negatively charged particles which Faraday had called ions. According to this theory, the solutions were capable of conducting electricity.

(From "Svante Arrhenius", Wikipedia)

Thanks to Arrhenius' theory of electrolytic dissociation, a number of difficult to explain chemical and physical phenomena and laws could be easily explained and described in a simple and understandable manner. Even though his theory has been subsequently modified in the following decades, it still remains a major discovery in chemistry. Arrhenius also made original contributions to many other aspects of chemistry and physics, "Green house effect" being one of them.

Early years

Svante Arrhenius was born at Vik near Uppsala on 19th February, 1859. His father Svante Gustaf Arrhenius worked as a land surveyor at the University of Uppsala and his mother Carolina Christina Thunberg was a homemaker. Svante Gustaf Arrhenius rose in the ranks to be in charge of the university's estates at Vik. When Arrhenius was barely a year old, his family moved to Uppsala proper from Vik. Even though his parents did not show any interest in skills of young Arrhenius, he learnt to read on his own. His father's work as a land surveyor required adding numbers and young Arrhenius was fascinated watching his father doing this in his account books. He became a child prodigy with numbers. In the process of watching his father at work, he developed prodigious arithmetical skill and an exceptional pictorial memory. This fascination with numbers remained throughout his life. Later in his scientific work, Arrhenius used these skills to discover mathematical relationships and deduce laws.

Arrhenius started formal schooling at the well known Cathedral School, Uppsala, when he was 8 years old joining the fifth grade straight away. He was motivated by the Rector of the school who was a good physics teacher. After graduating from high school with distinction in physics and mathematics, he joined the University of Uppsala in 1876 to study physics, mathematics and chemistry. After obtaining a basic degree in chemistry in 1878, he enrolled for doctoral work in physics but he was not happy with either his chief supervisor of physics (Tobias Thalen, a fairly recognised physicist for his analysis of spectra) or with his chemistry supervisor (Per Teodor Cleve, the only available guide in chemistry). They were rigid in their approach and did not appreciate his ability to think out of the box and arrive at apparently novel ideas on vexing problems. He decided to leave the University of Uppsala in 1881 after three years of frustrating graduate work. Fortunately, he was granted special permission to continue his doctoral work in absentia to work with Erik Edlund at the Physical

Institute of the Swedish Academy of Sciences in Stockholm. This move apparently antagonised the professors at the University of Uppsala.

Arrhenius conducted a number of experiments during 1882-83 to determine the conductivity of electrolytes and this study formed the subject of his doctoral thesis *Recherches sur la conductibilité galvanique des électrolytes* (Investigations on the galvanic conductivity of electrolytes). His dissertation (running to many pages) contained in addition to detailed description of his experiments on conductivity of electrolytes, 56 hypotheses including his radical theory of electrolytes. He hypothesized that when dissolved in a solvent like water, molecules of certain acids, bases, and salts dissociate into positive and negative ions. (*Théorie chimique des electrolytes).* This idea was revolutionary as the prevalent idea was Michael Faraday's (and others) that ions were produced only when electric current was passed through the solution of certain compounds and decomposed the substance. According to Arrhenius, the nature of the substance and its concentration in solution decided the degree of dissociation. He added further that ions not only conducted electricity but were responsible for the chemical reaction in solutions that took place between these oppositely charged ions.

Arrhenius's explanation that while neither pure salts nor pure water can conduct electricity, solutions of salts in water are conductors of electricity was the most important idea of the dissertation. His experiments with sugar and salt in water to highlight the difference between electrolytes and non-electrolytes were ingenious. He observed their solutions behaved differently with respect to lowering of the freezing point of water and the passage of electricity through them. Sugar solution did not allow electricity to pass through it while salt solution did. He noticed that while both of them brought down the freezing point of water below 0°C, in the case of sugar solution the lowering of the freezing point was proportionate to the number of sugar molecules added but in the case of salt solution, the lowering of the freezing point was twice in proportion to the number of salt molecules added. After repeating the experiment many times, he concluded that the reason for this in both cases was the same - the manner in which they dissolved in water. Sugar just dissolves in water i.e. their molecules do not break up on dissolving, salt molecules split into sodium ion and chlorine ion with opposite charges on dissolving in water. This made salt solution conduct electricity.

Berzelius, another eminent Swedish chemist had suggested that there was a relation between chemical reaction/affinity and electricity but unfortunately this had been so completely forgotten by scientists that the intrinsic value of Arrhenius's work was not understood by both his superiors and other contemporary scientists in Sweden.

The examining committee of the Ph D thesis of Arrhenius (which included Per Theodor Cleve) was not impressed by his idea of dissociation and after much discussion awarded a fourth class for the thesis (*non sine laude approbatur*, 'approved not without praise') and a third (*cum laude approbatur*, 'approved with praise') for his defense. This was a bitter disappointment to Arrhenius as this score was too low for Arrhenius to be eligible for either a research or a teaching position (docentship) and his dream of pursuing an academic career seemed to be at an end.

It is said that when Ostwald met Theodor Cleve on his visit to persuade Arrhenius to join him, Theodor Cleve apparently told Ostwald that it was difficult to accept potassium chloride separated to form potassium and chlorine in solution. At the banquet held in Stockholm in 1903 to felicitate Arrhenius winning the Nobel Prize, it is reported that Theodor Cleve paid him a left handed compliment by saying *"These new theories (of Arrhenius).... also suffered from the misfortune that nobody really knew where to place them. Chemists would not recognize them as chemistry; nor physicists as physics. They have in fact built a bridge between the two"* (From Daintith [2]).

The silver lining during those dark days was the highly positive review of the thesis by the professor of chemistry at the Technical High School of Stockholm, Sven Otto Pettersson, published in the journal *Nordisk Revy*. He wrote *"The faculty have awarded the mark 'non sine laude' to this thesis. This is a very cautious but very unfortunate choice. It is possible to make serious mistakes from pure cautiousness. There are chapters in Arrhenius' thesis which alone are worth more or less all the faculty can offer in the way of marks"* (From Encyclopedia.com[3]).

Fortunately for Arrhenius, he had sent copies of his dissertation to a number of leading physical chemists in Europe including van't Hoff, Wilhelm Ostwald, and Rudolf Clausius all of whom were impressed by his novel idea. In fact, Professor Ostwald was so impressed with Arrhenius's thesis that he called it the most important publication on the theory of affinity and went to Sweden to personally invite Arrhenius to join van't Hoff and himself in their ongoing research in physical chemistry. But Arrhenius refused the offer. Erik Edlund used his influence to get him a travelling fellowship from the Academy of Sciences which allowed him to work for several years with various eminent chemists in other parts of Europe. Between 1886 and 1888 he worked with Ostwald, Kohlrausch, Boltzmann and van't Hoff in various laboratories of Europe. van't Hoff had noticed certain solutions behaved as if there were more molecules than expected. Based on the theory of electrolytic dissociation, Arrhenius conducted experiments and was able to explain the rationale (which had not been entirely understood) for van't Hoff's equation for osmotic pressure. Based on the data obtained from his experiments, Arrhenius published in 1887 the classic paper *Über die Dissociation der in Wasser gelösten Stoffe* (On the Dissociation of Substances in Water). He also applied his dissociation theory to explain why the lowering of the freezing point and increase of the boiling point occurred in solutions containing electrolytes. His theory helped to explain many chemical phenomena such as the laws of vapor pressure lowering and freezing point depression, the dilution law of Ostwald. It also explained chemical reactions of electrolytes in aqueous solutions in quantitative terms. While this increased his reputation internationally, he was yet to get the recognition he so richly deserved in his own country.

Ironically, years later in 1903 he was awarded the Nobel Prize in chemistry *"in recognition of the extraordinary services he has rendered to the advancement of chemistry by his electrolytic theory of dissociation"* (From Nobelprize.org [4]) - the very work that had caused him so much hardship.

Based on his theory of dissociation, Arrhenius also arrived at his own definitions for acids and bases. According to him acids were substances that contained hydrogen and yielded hydrogen ions in aqueous solution with water as the solvent and bases were substances that contained OH group and yielded hydroxide ions in aqueous solution with water as the solvent.

Career of Arrhenius

Arrhenius had a long and distinguished career in both physics and chemistry and it can be viewed as three distinct periods. Interestingly, as he changed his research interest, he also changed his institution and he was at three different institutions during these three periods. He was at the Physical Institute of the Academy of Sciences in Stockholm and at foreign universities as a doctoral and post-doctoral research fellow between 1884 and 1890 when he did path-breaking work on dissociation of electrolytes. In 1891, he accepted a lectureship at the Stockholm's Högskola (presently known as the University of Stockholm) and became a professor of physics in 1895. He carried out his work on the role of carbon dioxide in the rising of atmospheric temperature among other things in cosmic physics and finally between 1901–07 he worked on the chemistry of toxins and antitoxins at the State Serum Institute in Copenhagen and the Nobel Institute for Physical Chemistry which was established in 1905 in Stockholm.

"Physical chemistry is the child of these two sciences (physics and chemistry); it has inherited the extensive character from chemistry. Upon this depends its all-embracing feature, which has attracted so great admiration. But on the other hand it has its profound quantitative character from the science of physics" (From Arrhenius [1]).

"The theoretical side of physical chemistry is and will probably remain the dominant one; it is by this peculiarity that it has exerted such a great influence upon the neighboring sciences, pure and applied, and on this ground physical chemistry may be regarded as an excellent school of exact reasoning for all students of the natural sciences" (From Arrhenius [1]).

On joining the Physical Institute of the Swedish Academy of Sciences Arrhenius at first was assisting Erik Edlund, but soon he was allowed to work on his own. His most important contribution to chemistry was also the result of his first independent idea - idea of electrolytic dissociation.

"I was led to the conclusion that at the most extreme dilutions all salts would consist of simple conducting molecules. But the conducting molecules are, according to the hypothesis of Clausius and Williamson, dissociated; hence at extreme dilutions all salt molecules are completely disassociated. The degree of dissociation can be simply found on this assumption by taking the ratio of the molecular conductivity of the solution in question to the molecular conductivity at the most extreme dilution" (From Partington [5]).

Arrhenius Equation

During numerous experiments Arrhenius had conducted, he had noticed that most chemical reactions required additional heat energy to proceed further i.e. addition of heat energy or rise in temperature influenced the rate of many chemical reactions. While collaborating with van't Hoff and Ostwald, Arrhenius worked further on the equation proposed by van't Hoff in 1884 where he used it to illustrate the dependence of equilibrium constants in a reversible reaction on temperature. Based on the experimental data, Arrhenius arrived at the concept of activation energy - the energy barrier that the two reacting molecules must overcome before the reaction can start. In 1889, Arrhenius published the famous equation $k = Ae^{-Ea/(RT)}$, where k is the rate constant of a chemical reaction, T, the absolute Temperature in Kelvin, A, a prefactor, E_a, the activation energy and R, the Universal gas constant, to express the relation between the rate constant of a chemical reaction and temperature. This equation gave a simple interpretation of van't Hoff's equation. The unique feature of the Arrhenius equation is that it provides a quantitative value to the relation between activation energy and the rate at which the

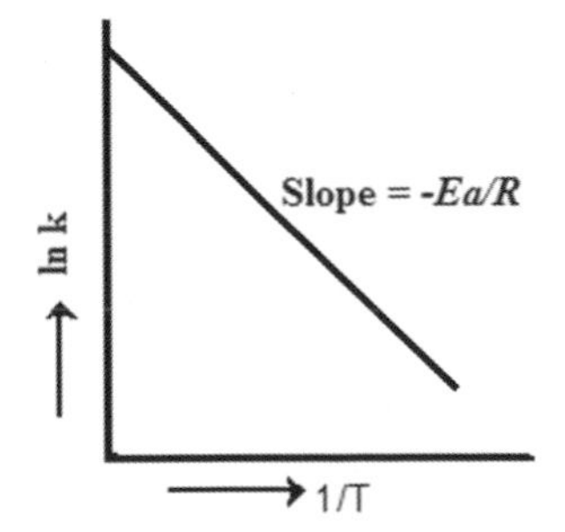

chemical reaction proceeds. It also supports the widely accepted generalization that in most common chemical reactions, increase of temperature by 10^0 Celsius doubles the rate of reaction.

The great insight of Arrhenius into the phenomenon of temperature dependence on the rate of a chemical reaction was the recognition of the formation of activated molecules from the reactant molecules as being the crucial step in any chemical reaction and also his suggestion that both of them were in equilibrium state with only the activation energy separating them.

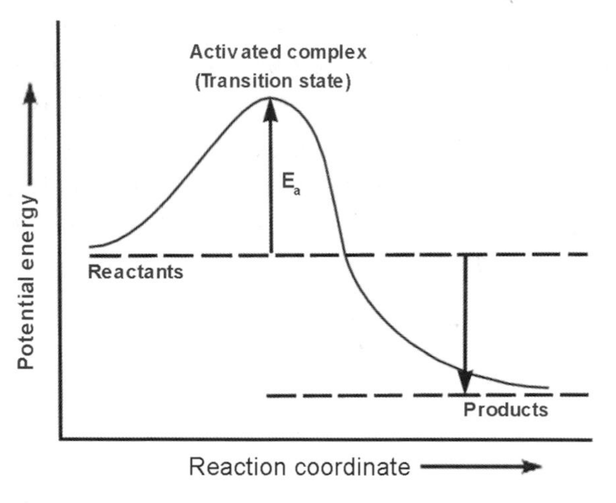

Based on this explanation, increasing the temperature resulted in the formation of more activated molecules to undergo the reaction.

> Ostwald used the above concept of Arrhenius in arriving at his theory of catalysis. He showed experimentally that the presence of a catalyst in a chemical reaction required smaller activation energy than the reaction without the catalyst.

Cosmic physics

Gradually, Arrhenius' theories were accepted by scientists in Europe. In 1891, Arrhenius joined *Stockholms Högskola* (later renamed Stockholm University) as a lecturer and was promoted against much opposition as professor of physics within four years (1895). He became rector in 1896.

On joining the *Stockholms Högskola*, Arrhenius changed his research interests by initially focusing on what he and his colleagues in Stockholm

University called cosmic physics which included the physics of the atmosphere, the oceans and seas and the landmasses. They set out to develop theories to explain the interdependence of the natural phenomena of these three distinct spheres. Arrhenius soon found that one of the important issues that preoccupied the scientists at that time related to *"the probable causes of the Ice Age"* (From Arrhenius [6]). He became fascinated by the role of carbon dioxide in the variations of air temperature. The crucial question, *"is the mean temperature of the ground in any way influenced by the presence of heat-absorbing gases in the atmosphere?"*, (From Arrhenius [6]) had intrigued physicists for a long time. Arrhenius wanted to find out what triggered the major changes or fluctuations in earth's temperature spanning geological eras and wondered if the culprit was the changes in the levels of carbonic acid (carbon dioxide forming the acid with water vapor) in the atmosphere. *"Conversations with my friend and colleague Professor Högbom, together with the discussions .... led me to make a preliminary estimate of the probable effect of a variation of the atmospheric carbonic acid on the temperature of the earth. As this estimation led to the belief that one might in this way probably find an explanation for temperature variations of 5°-10°C., I worked out the calculation more in detail, and lay it now before the public and the critics"* (From Arrhenius [6]).

Arrhenius applied the principles of physical chemistry to successfully study meteorology, cosmology, and biochemistry. In meteorology, he concentrated on the cause of ice ages and as a spin off, he discovered the role of CO_2 in causing the "the greenhouse effect" in the atmosphere and thus anticipated the late 20th century preoccupation with global warming and climate change.

Arrhenius used Joseph Fourier's study that showed how atmosphere acted like a sheet of glass in a greenhouse that allowed the incoming short wave solar radiation to reach the earth's surface but did not allow the long wave terrestrial radiation given out by the earth's heated surface (after sunset) to escape, thus trapping both insolation and terrestrial radiation. He made use of the results of the work done by Samuel Pierpont Langley and Frank Washington Very at Allegheny Observatory, Pittsburgh (who calculated the quantity of infrared radiation from the moon). After meticulous calculations he was able to predict correctly how the rise in the level of carbon dioxide in the atmosphere and the rise in temperature of the lower levels of the earth's atmosphere were directly related and that if the level of atmospheric carbon dioxide increased in geometric progression, it would result in the temperature of the earth's atmosphere increase in arithmetic progression.

> While Joseph Fourier is credited with coining the term Greenhouse effect, Arrhenius had the insight to see the relation between increase in the levels of carbonic acid and the warming of the earth's atmosphere. Arrhenius calculated the result of the greenhouse effect and declared that *"if the quantity of carbonic acid [CO_2] increases in geometric progression, the augmentation of the temperature will increase nearly in arithmetic progression"* (From "Svante Arrhenius", *Wikipedia* [7]). This is often referred to as Arrhenius' Greenhouse Law. A modified version of this calculation is used even today.

Arrhenius used his model of Greenhouse effect to explain the cause for the ice age phenomenon. It occurred to Arrhenius that human activities such as burning of fossil fuels and other combustion activities also contributed greatly to the increase in atmospheric carbon dioxide beyond acceptable levels and that this would cause the reverse phenomenon (an increase in temperature). He became the first scientist to predict that excessive emission of carbon dioxide would result in global warming. Arrhenius anticipated the rise in carbon dioxide levels to such dangerous levels to occur after 3000 years – a thought he found quite comfortable. He wrote in his book Worlds in the Making *"By the influence of the increasing percentage of carbonic acid in the atmosphere, we may hope to enjoy ages with more equable and better climates, especially as regards the colder regions of the earth, ages when the earth will bring forth more abundant crops than at present for the benefit of rapidly propagating mankind"* (From Arrhenius [8]).

He reassured the general public that there was little danger of the return of the Ice Age as emission of CO_2 by human activities would be strong enough to prevent the world from entering a new ice age.

"Since, now, warm ages have alternated with glacial periods, even after man appeared on the earth, we have to ask ourselves: Is it probable that we shall in the coming geological ages be visited by a new ice period that will drive us from our temperate countries into the hotter climates of Africa? There does not appear to be much ground for such an apprehension. The enormous combustion of coal by our industrial establishments suffices to increase the percentage of carbon dioxide in the air to a perceptible degree" (From Arrhenius [8]).

In addition to carbon dioxide, Arrhenius included the role of water vapor, latitude of a place as factors influencing the variations in the temperature of the lower layers of the earth's atmosphere. However, he did not include clouds and upward movement of heated air in the atmosphere as contributing factors - factors which are now considered crucial for accurate calculations.

Arrhenius demonstrated that to get the temperatures of the Arctic region similar to those that prevailed in the ice ages in 1896, the carbon dioxide levels had to decrease by 0.62-0.55 times; conversely, he also stated that there would be an increase by two to three times in the temperature of the Arctic region if there was an increase of 2.5 to 3 times in the level of carbon dioxide of 1896.

Arrhenius was the first scientist to undertake a serious study of the relationship between increase in the atmospheric carbon dioxide levels as a result of burning of fossil fuels and other CO_2 producing fuels resulting in global warming. By discovering the influence of CO_2 on the changes in the earth's temperature, he unraveled the secrets of the earth's atmosphere and triggered research into the causes of climate changes. For this unique insight into the negative effect of industrial and other negative human activities on global warming, he is considered by many climatologists as the father of climate change.

While he was probably correct in his forecast about the ice age, he was wrong about the time scale regarding the increase of carbonic acid in the atmosphere and the beneficial effect of such increase. However he gave us a warning about the over use of natural resources when he cautioned *"Humanity stands ... before a great problem of finding new raw materials and new sources of energy that shall never become exhausted. In the meantime we must not waste what we have, but must leave as much as possible for coming generations"* (From Arrhenius [9]).

Toxins and antitoxins

During 1901-1907, Arrhenius switched his research to the chemistry of toxins and antitoxins or immunochemistry. During this period, he worked in Copenhagen at the State Serum Institute and in Stockholm at the Nobel Institute for Physical Chemistry established in 1905. He applied the methods of physical chemistry to investigations into physiological problems and the study of toxins and antitoxins. He was able to establish that the same laws applied to both the reactions in a test tube and within living organisms. He gave a series of lectures at the University of California 1904 to show how the methods of physical chemistry could be used to study toxins and antitoxins. He published these lectures as a book titled *Immunochemistry* in 1907.

Karl Landsteiner has remarked *"Arrhenius has shown that the quantitative relationship between toxin and antitoxin is very similar to that between acid and base"* (From Landsteiner & Jagic [10]).

Popularising science

Arrhenius reached out to young people as well as to the general public by giving popular lectures and writing books to explore thought-provoking ideas. During the last few years of his life, he wrote a number of books which proved so popular that they went into many editions and were translated into several languages as well. Among them, the most popular one was *Världarnas utveckling* (Worlds in the Making) published in 1906, in which he hypothesized that bacteria and light energy were responsible for spreading life in the universe. *Stjärnornas Öden* (Destiny of the Stars) published in 1915, *Smittkopporna och deras bekämpande* (Smallpox and its combating) in 1913 and *Kemien och det moderna livet* (Chemistry and modern Life) in 1919 were other popular books. Arrhenius also contributed an article on physical chemistry for the 13th edition (1926) of the *Encyclopædia Britannica.*

In addition to popular books, Arrhenius published books on various topics in both physics and chemistry during a period spanning close to two decades. Some of them are: Textbook of Theoretical Electrochemistry (1902), Textbook of Cosmic Physics (1903), Theories of Chemistry (1907), Worlds in the Making: The Evolution of the Universe (1906), Immunochemistry and The Life of the Universe as Conceived by Man from the Earliest Ages to the Present Time (1907), Theories of solutions (1912), Quantitative Laws in Biological Chemistry (1915), and Chemistry and Modern Life (1919).

Honours

Arrhenius' outstanding contributions to both physics and chemistry was recognized world over. He was elected to the Swedish Royal Academy of Sciences against much opposition in 1896 and was finally accepted in Sweden after he won the Nobel Prize. In 1901, the Royal Swedish Academy of Sciences elected him as a member. Around 1900, he got involved in setting up the Nobel Foundation to award the Nobel Prizes. He was a member of the Nobel Committee on Physics to select the annual Prize winners. Even though he was not a member of the Chemistry committee, he was a de facto member with enormous powers. It is said that he was so powerful that he decided on not only who should get the prize but also who should not.

Arrhenius was elected Foreign member of the Royal Society (1911), awarded Davy medal of the Royal Society London (1902), the Willard Gibbs Award (1911), the Faraday Medal (1914), Foreign Honorary Member of the American Academy of Arts and Sciences (1912), Franklin Medal from USA (1920).

He received Honoris Causa degrees from Universities of Cambridge, Oxford. Edinburgh, Birmingham, and Leipzig, Heidelberg, Greifswald and Groningen.

Final days

Svante Arrhenius had a serious setback to his health when he came down with severe inflammation of the intestinal mucous membranes in September 1927. He did not recover from this and succumbed to it within a month. He died on October 2, 1927 in Stockholm. He was only 68 years old at that time.

References

1. Svante August Arrhenius, *Theories of solutions*, Yale University Press, 1912.
2. *Biographical Encyclopedia of Scientists, Third Edition,* edited by John Daintith, CRC Press, Taylor & Francis Group, p27, 2008.
3. *Nordisk revy* (15 December 1884); *cf. Svensk kemisk tidskrift* (1903), 208. (From "Arrhenius, Svante August." Complete Dictionary of Scientific Biography. 2008. Encyclopedia.com. (March 30, 2015). http://www.encyclopedia.com/doc/1G2-2830900169.html).
4. The Nobel Prize in Chemistry 1903, *Nobelprize.org.* Nobel Media AB 2014. Web. 30 Mar 2015. http://www.nobelprize.org/nobel_prizes/chemistry/laureates/1903/
5. J. R. Partington, *A History of Chemistry* (1961), Vol. 4. & Chemical Society, *Memorial Lectures delivered before the chemical society*, Volume 3, Pennsylvania State University, 1933.
6. Svante Arrhenius, "On the Influence of Carbonic Acid in the Air Upon the Temperature of the Ground," Philosophical Magazine 1896(41): 237-76 (1896).
7. "Svante Arrhenius." *Wikipedia, The Free Encyclopedia*. Wikipedia, The Free Encyclopedia, 27 Mar. 2015. Web. 30 Mar. 2015.
8. Svante Arrhenius, Worlds in the making: The Evolution of the Universe, Trans. by Dr. H. Borns, Harper & Bros., New York, 1908.
9. Svante Arrhenius, *Chemistry in modern life*, Translated by Clifford Shattuck Leonard, D. Van Nostrand company, 1925, Digitized 14 Jul 2007.
10. Landsteiner and Nicholas von Jagic, 'Uber Reaktionen anorganischer Kolloide und Immunkorper', *Münchener medizinischer Wochenschrift* (1904), **51**, 1185-1189. Trans. Pauline M. H. Mazumdar (from the website, (From Today in Science History, Dictionary of Science Quotations, Svante Arrhenius Quotes).

13. ALFRED WERNER (1866–1919)

Inorganic Kekule

(From "Alfred Werner", Wikipedia)

"Occasionally, a single man may play such a central role in a particular field of science that his name becomes synonymous with that field. Alfred Werner the undisputed founder and systematizer of coordination chemistry is just such a man. Even today, almost a full half-century after his death, coordination compounds, particularly metal-ammines, are still known as Werner complexes, and the coordination theory is colloquially called Werner's theory" (From Kauffman [1]).

"Werner's fame is ..... grounded in inorganic chemistry. He began with a study of metal-ammines, hydrates, and double salts; but his ideas soon encompassed almost the whole of systematic inorganic

chemistry and even found application in organic chemistry. He was the first to show that stereochemistry is a general phenomenon and is not limited to carbon compounds, and his views of valence and chemical bonding stimulated subsequent research on these fundamental topics" (From Encyclopedia.com [2]).

"(Werner's) coordination theory, with its concepts of coordination number, primary and secondary valence, ... octahedral, square planar, and tetrahedral configurations, not only provided a logical explanation for known "molecular compounds" but also predicted series of unknown compounds. ......Werner recognized and named many types of inorganic isomerism.... He also postulated explanations for polynuclear complexes, hydrated metal ions, hydrolysis, and acids and bases" (From Encyclopedia.com [2]).

Early years

Alfred Werner was born on December 12, 1866 in Mulhouse, Alsace (France) to J. A. Werner who worked in a foundry and Salomé Jeanette, a homemaker. His father was a locksmith. As he was the youngest of the four children, he enjoyed extra attention from his family. His family remained in Mulhouse even after the Alsace region was annexed by Germany in 1871.

Werner started his schooling in Alsace at the École Libre des Frères in 1872 at the age of six years. Even at this young age his basic traits of supreme self confidence and refusal to submit blindly to authority were evident and got him into trouble with the strict discipline at school. Werner joined the École Professionelle (a technical school) in Alsace in 1878 and developed interest in chemistry while he was there. He even built a small laboratory in the barn at the back of his house. Much like Humphry Davy, he confidently submitted a manuscript ("Contribution de l'acide urique, des séries de la théobromine, caféine, et leurs derivés") to the director of the Mulhouse Chemie-Schule. The paper was unremarkable both for its chemistry and the style of presentation but was a forerunner of his bold and unconventional daring in attacking scientific problems. His interest in chemistry was further reinforced when he attended chemistry classes of Engler at the Technical High School in Karlsruhe during the compulsory military service (1885-1886). In 1886 he joined Eidgenössisches Polytechnikum (now called Eidgenössisches Technische Hochschule [ETH], or Swiss Federal Institute of Technology) in Zürich, Switzerland, from where he obtained a diploma in technical chemistry in 1889. He then worked under Professor Hantzsch at the Polytechnikum for his doctoral study in organic chemistry and received his doctorate in 1890 from the University of Zurich for his thesis titled, *"Über räumliche Anordnungen der Atome in stickstoffhaltigen Molekülen"* (From Kauffman [3]). This was one of the important papers he published between 1890 and 1893.

> Werner was overjoyed by his dissertation receiving much attention. He wrote to his parents on May 11, 1981 *"I am beginning to take my place among the chemists of the time, and if heaven preserves my health, I intend to surpass them one by one, for glory is not an empty word; it is the personal satisfaction of a man who needs it as a stimulant in his moments of weakness"* (From Kauffman [1]).

Werner was influenced by both French and German cultures. Although he had great admiration for German science and published most of his research in German journals, he was also influenced by the French political and cultural attitudes (especially the spirit of independence and rebellion against authority). Werner was unconventional in thinking. He extended the concept of the tetrahedral asymmetric carbon atom of van't Hoff and Le Bel to the asymmetric nitrogen atom with the valence of three and proposed the revolutionary idea that the nitrogen atom also has tetrahedral geometry with the atom's three valence bonds directed towards the three corners of a tetrahedron with nitrogen atom itself occupying the fourth corner of the tetrahedron. It was a path breaking concept as it established the tetrahedral stereochemistry of trivalent nitrogen almost on par with that of tetravalent carbon and explained the reason for *cis-trans* isomerism in many nitrogen compounds.

After obtaining his doctorate degree, Werner wanted to pursue an academic career but could not get an opening immediately. He continued his research for the next three years at ETH and at the Collège de France, Paris, under Berthelot on thermochemistry. Werner also decided to work on his *habilitationsschrift*, (research performance required to obtain a teaching position at a university) and chose "Beiträge zur Theorie der Affinität und Valenz" (From Kauffman [3]) as the topic in which he challenged Kekule's concept of fixed valence and the classification of compounds as valence compounds (compounds which had fixed valence) and molecular compounds (compounds which did not have fixed valence). Werner proposed a more flexible concept with affinity as the force of attraction exerted by an atom from its centre and directed uniformly towards the atom's surface. As the paper was published in a relatively unknown journal, this important work went unnoticed by the wider scientific community. His *habilitationsschrift* was accepted by the Swiss authorities and Werner returned to the Polytechnikum at Zurich as a *Privatdozent.*

The laboratory at the Polytechnikum was small with insufficient lighting and inadequate ventilation. Its nickname among students was catacombs (burial chambers of early christians), but Werner did his path breaking work in there.

From organic to inorganic chemistry

Even though Werner started his professional career as an organic chemist, he was becoming more and more interested in conducting research on inorganic compounds. Today, he is remembered as an extraordinary inorganic chemist. His gift for thinking in three dimensional space and his training in organic stereochemistry served him well in his research on compounds in which ammonia or amines were bonded to metal atoms at the centre of the compound. In 1893, he published his seminal paper in inorganic chemistry "Beitrag zur Konstitution anorganischer Verbindungen," in *zeitschrifit für anorganische Chemie* (From Encyclopedia.com [2]). He moved to the University of Zurich as an Associate Professor the same year and became a full professor two years later. He remained there for the rest of his life. Werner married Emma Wilhelmina Giesker in 1894 and became a Swiss citizen.

Coordination theory

Much like Kekule's dream inspiring him to solve the structure of benzene, Werner, then a young and rather unknown *Privatdozent* working on the puzzle of 'molecular compounds', seems to have got the answer to this puzzle in a dream at 2 am in 1892. Werner reportedly shot out of bed and wrote the entire seminal paper "Beitrag zur Konstitution anorganischer Verbindungen" without pausing until he finished it at 5 pm, explaining his theory of coordination compounds.

Before Werner, valence of an element was defined based on the number of bonds formed by it. Werner proposed the unique concept of two types of valence: hauptvalenz and nebenvalenz. He called hauptvalenz as the primary valence and nebenvalenz as the secondary valence. He postulated that in a coordination compound, the central atom (usually a transition metal) has two types of valences. According to his theory, the primary valence can be normally ionized and is satisfied by negative ions. It is represented by the oxidation number of the central atom. Primary valences are non-directional. The secondary valence is a weaker valence and cannot be ionized. It is satisfied by both negative ions and neutral molecules. The number of ions or molecules bonded to the central atom is called the coordination number and the secondary valence is represented by the coordination number. Coordination numbers can be 3, 4, 6, 8. Secondary valences are directed towards fixed positions in space to form a definite geometric shape depending on the coordination number (tetrahedral when the coordination number is four and octahedral when it is six). For example, in the compound, $[Co(NH_3)_6]Cl_3$, the Co is the central transition metal atom, with a primary valence of 3 and the NH_3 molecules within the brackets bonded to the central Co atom are from the secondary valence The secondary valence of 6 is the coordination number.

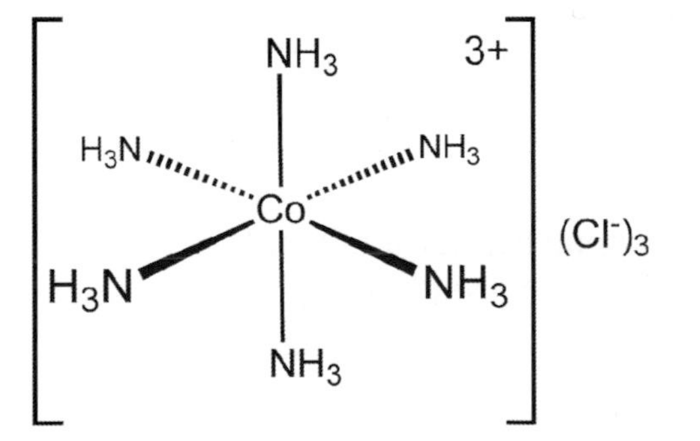

When Werner's theory was published in 1893, leading chemists refused to accept it as the theory was not supported by experimental evidence. Werner had made use of the data collected by other chemists. Sophus Mads Jørgensen of Denmark, supported Kekule's chain bonding and was the most outspoken critic of Werner.

Werner's theory of coordination was profound from two view points: a) it helped to solve structures of known coordination compounds and the existence of isomers. b) it was able to predict the existence of innumerable unknown compounds and their isomers. He had predicted the existence of ammonia-violeo salts but he and his co-workers had to work single-mindedly for close to 25 years preparing thousands of samples before they succeeded in preparing a stable ammonia-violeo salt $[Co(NH_3)_4Cl_2]Cl$ in 1907. It is said that Werner was so excited by this that he immediately wrote about his success and sent Jørgensen a sample of the compound by mail even before he submitted the manuscript for publication. Jørgensen accepted this as proof of Werner's theory of coordination compounds.

Werner had to experimentally resolve a coordination compound with optical activity to support his theory. In 1911, Werner and his student V L King succeeded in resolving the cobalt compound they had prepared in the laboratory into cis and trans isomers. King recalls how when he excitedly broke the news to Werner, *"he just leaned back in his chair smiled and said not a single word"* (From King [4]). Was it because he knew that he would be proven right in the end? Werner rushed to the laboratory and worked with King till late in the night preparing many derivatives of the compound and measuring their optical activity. When they found that the compound retained its optical activity in solution even when heated to near boiling temperature or left standing for a long period, it became clear that they had succeeded in resolving a coordination compound with octahedral configuration for the first time. Prior to this, chemists had succeeded in resolving compounds of asymmetric atoms with tetrahedral configurations. Werner considered this success as a *"stereochemical achievement of the first order"* and wanted to share his excitement with everyone (From Read [5]). This achievement *"shook chemistry to its innermost foundations"* (From Kauffman [6]) *and* according to Morgan *"The spatial configuration of coordination complex with six units is now as firmly established as that of the tetrahedral carbon atom"* (From Morgan [7]).

Werner considered the resolution of cobalt compound as one of his most profound contributions as it confirmed the octahedral geometry that he had predicted. His experiment *"proved that metal atoms can act as central atoms of stable and asymmetrically constructed molecules (and) that pure molecular compounds (coordination compounds) can also occur as stable mirror image isomers, whereby the difference between valence compounds (ordinary compounds), which is still frequently maintained, completely disappears"* (From Kauffman [8]) According to King, *"it provided the last proof for the octahedral formula assumed by Werner"* (From Kauffman [9]).

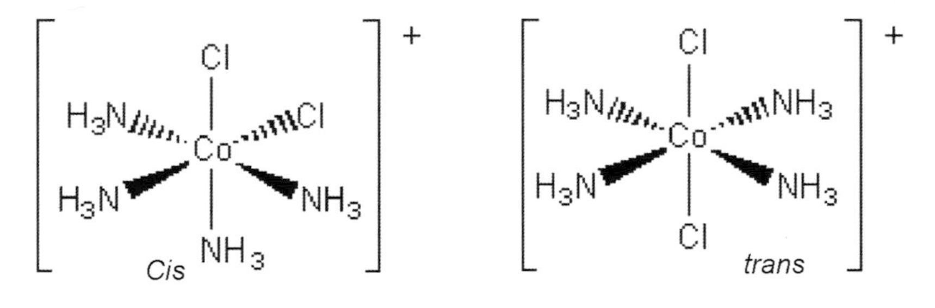

The day the resolution of the cobalt compound was achieved, students were nonplussed when Werner (who was never late for his lectures) did not come for his 5 pm lecture. A few minutes later, one of his young students entered the lecture hall and announced the cancellation of the lecture. At once the entire class knew something extraordinary had been achieved in the laboratory. That december for the traditional in-house Christmas party of the laboratory, as a tribute to the monumental accomplishment, Paul Karrer (himself a future Nobel Laureate) a young student at that time, working with Alfred Werner, wrote an irreverent play in verse "*Drehen und Spalten*" (Rotating and Resolving), narrating the story of the resolution of an optically active coordination compound for the first time in which der alte (the old one) Werner says *"Give me the Cobalt in a trice, Though Hell and Devil may say nay, I shall resolve it today"* (From Kauffman [10]). An international cast of chemists gathered for the Le Bel-van't Hoff Centennial symposium of the 168th meeting of the American Chemical Society at Atlantic City, September 11-12, 1974 performed the play.

"Whenever Werner opened up a new field, he expanded (it) with unbelievable speed" (From Kauffman [6]). Once the breakthrough was achieved, (as if to make up for the years spent on this problem) Werner resolved a number of coordination compounds of cobalt whose existence his theory had predicted and published more than 40 papers in eight years. However, he still had to experimentally prove the existence of optical isomerism in coordination compounds without a carbon atom in them. Finally, in 1914, he succeeded in preparing a carbon free asymmetric cobalt compound and he published a paper in *Ber Dtsch. Chem. Ges* (From Kauffman [11]) in which he said that *"carbon free inorganic compounds can also exist as mirror image optical isomers and the difference still existing between carbon compounds and purely inorganic compounds disappears"* (From Kauffman [11, 8]). With this experimental proof, finally even the most diehard critics had to accept Werner's conviction that mirror image optical isomerism exists in certain specific group of carbon free inorganic compounds.

Kekule used the asterisk (*) notation in what he called the molecular compounds. For example, $[Co(NH_3)_6]Cl_3$, which Kekule considered as a molecular compound, was written as $CoCl_3*6NH_3$. While this assisted chemists to fix the valence of the central metal atom, it did not give any clue about the structure and shape of the coordination compound. Werner's theory solved the puzzle by proposing two valences for the central transition metal atom.

Periodic Table, invaluable toolkit

What is remarkable about Werner's achievement is that he was able to formulate his theory without any of the modern analytical tools such as

UV, IR, NMR and EPR spectroscopies or X-Ray crystallography. Werner made use of traditional tools such as smell, taste, melting point and conductivity and polarimetry. Periodic table was his invaluable "toolkit". It provided Werner all the information that he needed to identify compounds of transition elements (particularly of platinum, cobalt) with different hauptvalenz (primary valence) and nebenvalenz (secondary valence). In fact, definite identification of existing elements and isolation of hitherto unknown elements helped Werner to publish in 1905 a modified version of the periodic table which included 16 more elements. In this version, lanthanide elements or rare earth elements (atomic numbers 58-71), were moved to a separate column as found in the modern periodic table.

Werner's revolutionary contributions gained acceptance and had an impact on almost the entire field of inorganic chemistry. He was one of the earliest chemists to prove stereochemistry was a general phenomenon and not a phenomenon limited to organic chemistry alone. Though he published a number of papers setting forth his theory of coordination compounds, his theory became famous through his two books *Lehrbuch der Stereochemie* (1904) and the monumental *Neuere Anschauungen auf dem Gebiete der anorganischen Chemie* (1905). The latter book became so popular that it went through several reprints and was translated into many languages. Even today it is considered to be the bible of inorganic chemistry.

Many chemists call him "the inorganic Kekule" as his impact on inorganic chemistry was as profound as that of Kekule, Le Bel, van't Hoff and Couper on organic chemistry.

Honours

Werner was honoured by scientific bodies of various countries for his work. He was nominated five times for the Nobel Prize starting from 1908, but he had to wait until he established the experimental proof for his theory of bonding in inorganic compounds and their resultant geometry. He learned of his Nobel Prize for 1913 from the telegram *"Nobel Prize for chemistry awarded to you. Letter follows. — Aurivillius",* he received from the Royal Swedish Academy of Sciences. His reaction to the news was *"I had not completely eliminated the thought that it would come some day, but I hadn't expected it this year"* (From Kauffman [1]). Werner was the first inorganic chemist and first Swiss scientist to be awarded the Nobel Prize in Chemistry *"in recognition of his work on the linkage of atoms in molecules by which he has thrown new light on earlier investigations and opened up new fields of research especially in inorganic chemistry"* (From Nobelprize.org [12]).

It took another sixty years for inorganic chemists to get the Nobel Prize. The next Nobel Prize in inorganic chemistry was awarded to Ernst Fischer and Geoffrey Wilkinson in 1973. William N. Lipscomb received the Nobel Prize in 1976 and Henry Taube in 1983.

Werner the man

Werner had a larger than life personality and was a man who lived life to the full. He was a man in a hurry to accomplish his mission. Everyday, he was the first one to reach the laboratory and the last one to leave it. He worked with the same energy for six and half days a week (taking off only Sunday afternoons) and a short summer vacation as a break. He needed only a short nap in thc afternoon at his work place to refresh himself and resume his work with breakneck intensity. On the mornings of his lecture, he was the first to arrive even when he partied the previous night. He enjoyed the company of his students and coworkers. He loved good food and wine. He was a brilliant and inspirational teacher. This together with his informal and easy style of dealing with students inspired many of his students to take up research in chemistry as a profession. Yet he was an enigma. He was considered to be *"an unapproachable celebrity known to all yet really known by none. …… a sphinx — a towering, unshakable, inscrutable figure cloaked in mystery"* (From Kauffman [1]). Werner was a hard task master and he did not spare himself either from the demanding schedule. His unrelenting efforts to provide conclusive proof of the octahedral geometry of cobalt compound is a typical example of Werner's unshakeable belief in his ideas and his never say die attitude to provide experimental proof in spite of almost insurmountable difficulties faced by him and his team of workers.

Werner started as teacher of organic chemistry at the University of Zurich and it was only in 1902 -1903 that he was entrusted with the responsibility of teaching the main inorganic chemistry course. It is said that his lectures were so popular that no lecture hall of the University was big enough to accommodate the ever increasing number of students wanting to attend his lectures. After the publication of his second book *Neuere Anschauungen auf dem Gebiete der anorganischen Chemie* (1905) he had offers of professorship from many universities in Europe and in the USA. Werner refused all the offers preferring to stay in his beloved Zurich.

The last days

His uninterrupted hectic schedule of work over decades and excesses in eating and drinking had a disastrous effect on his health. His degenerative illness of arteriosclerosis (of the arteries to the brain) gave enough warnings of its onset but Werner was too obsessively preoccupied with his research to take the warning signs of chronic headaches and repeated inability to recall things. Things came to a head at the zenith of his triumph of receiving the Nobel Prize in 1913. Four years later, by the winter semester of 1917, Werner's headaches accompanied by giddiness and memory loss lasting a few minutes had become frequent and speech difficulties became noticeable. In a letter written on 24th August 1918 to the university, Werner reluctantly applied for leave for the winter semester, hoping to recover and resume his duties in the summer of 1919. Alas it was not to be. His health deteriorated rapidly and his wife informed the university on 6th May 1919 that his health had deteriorated so rapidly that he could not return to the university. Werner officially retired from the university to which he had brought so much fame and glory on 15th October, 1919 and died exactly a month later on 15th November 1919.

References

1. George B. Kauffman, *Alfred Werner, Founder of Coordination Chemistry*, Springer Berlin Heidelberg, 1966.
2. "Werner, Alfred." Complete Dictionary of Scientific Biography. 2008. *Encyclopedia.com.* 17 Apr. 2015 <http://www.encyclopedia.com>.

3. George B. Kauffman, *Foundation of nitrogen stereochemistry: Alfred Werner's inaugural dissertation, J. Chem. Educ.*, 1966, 43 (3), p 155-156.
4. V. L. King, *"A Rough but brilliant diamond,"* J. Chem. Educ., 1942, 19, 345.
5. J. Read, *Humour and humanism in Chemistry*, G. Bell & Sons, London 1947, p 264.
6. George B. Kauffman, *A stereochemical achievement of the first order: Alfred Werner's resolution of cobalt complexes, 85 years later,* Bull. Hist. Chem. 20 (1977), p50-59. (Quoted from P. Karrer, "Prof. Alfred Werner," Neue Zürcher Zeitung, No. 1804, Erstes Morgenblatt, November 21, 1919, Feuilleton p. 1).
7. G. T. Morgan, "*Alfred Werner*", J. Chem. Soc., 1920, 117, 1639-1648.
8. G. B. Kauffman; *Classics in Coordination Chemistry, Part 1, The selected papers of Alfred Werner*, Dover, New York, 1968, p159-173.
9. George B. Kauffman, *A stereochemical achievement of the first order: Alfred Werner's resolution of cobalt complexes, 85 years later,* Bull. Hist. Chem. 20 (1977), p50-59. (Quoted from V. L. King., "Uber Spaltungsmethoden und ihre Anwendung auf komplexe Metall Ammoniakverbindungen," Dissertation, Universitat Zurich, J. J. Meier, Zurich, 1912, p59).
10. George B. Kauffman, *A stereochemical achievement of the first order: Alfred Werner's resolution of cobalt complexes, 85 years later,* Bull. Hist. Chem. 20 (1977), p50-59. (Quoted from G. B. Kauffman and H. K. Doswald, "*PaulKarrer—Rotating and Resolving: A Tragicomic Popular Play,*" Chemistry, 1974, 47(2), 8-12).
11. George B. Kauffman, *A stereochemical achievement of the first order: Alfred Werner's resolution of cobalt complexes, 85 years later,* Bull. Hist. Chem. 20 (1977), p50-59. (Quoted from A. Werner, "Zur Kenntnis des asymmetrischen Kobaltatoms. XII. Ober optische Aktivität bei kohlenstofffreien Verbindungen," Ber. Dtsch. Chem. Ges., 1914, 47, 3087-3094).
12. "The Nobel Prize in Chemistry 1913". *Nobelprize.org.* Nobel Media AB 2014. Web. 20 Apr 2015. <http://www.nobelprize.org/nobel_prizes/chemistry/laureates/1913/>

14. RICHARD WILLSTÄTTER (1872–1942)

Ecstasy and Agony of Willstätter

Early years

Willstätter was born on August 13, 1872 in Karlsruhe, Germany, to Maxwell Willstätter, a textile merchant, and Sophie Ulmann Willstätter, who also came from a family in the textile business. As was the norm at that time, young Willstätter started his education in the Gymnasium at Karlsruhe. When he was eleven years old, there was a dramatic change in his life. His father decided to move to New York for a brief period to have better economic prospects and also get away from the covert anti-Semitism. This brief period stretched to more than a decade! Willstätter's mother, forced to manage her two young sons on her own moved with them to Nürnberg to be close to her family. Willstätter joined the technical school there, but young Willstätter found it difficult to adjust to the new surroundings where anti-Semitism was quite overt. As a result, he did not do well in Latin, the most important subject in the gymnasium even though he fared reasonably well in other subjects. His family decided that he should train for business instead of an academic career. Fortunately for young Willstätter and certainly for chemistry, before this plan could be implemented, inspired by teachers and excited by some of the experiments he conducted on his own at home, he started to do well at the gymnasium.

(From Nobelprize.org)

Although his first priority was to study medicine, his family circumstances could not afford the long years of study. Willstätter decided to become a chemist but he continued to be interested in biological processes. Later, Willstätter observed in his autobiography that excellence in academic subjects caused one to be disliked, while athletic excellence resulted in popularity.

Willstätter joined the University of Munich in 1890, aged 18, to study chemistry and obtained the basic degree in chemistry in 1893. He registered for his Ph.D there and began his research under Alfred Einhorn on the chemistry of cocaine and obtained his doctorate degree a year later in 1894. While working for his degree, he was noticed by Adolf von Baeyer, a famous organic chemist of that time. Willstätter considered himself a student of Baeyer even though he did not directly work under him. According to Willstätter, Baeyer was more *"of an encyclopedist than a researcher"* (From Willstätter [1]). He enjoyed watching Baeyer conduct difficult experiments using just a test tube. *"I watched Baeyer activating magnesium with iodine for a difficult Grignard reaction; it was done in a test tube, which he watched carefully as he moved it gently by hand over a flame for three quarters of an hour. The test tube was the apparatus to Baeyer"* (From Willstätter [1]). Working independently, Willstätter extended his research to other alkaloids. After two years, he became an unpaid lecturer in 1896 and a Professor extraordinarius (professor without a chair) in 1902.

Cocaine

During his doctoral research into the structure of cocaine, Willstätter did not agree with the structure of cocaine that Einhorn and his coworkers had arrived at. Not surprisingly, Einhorn did not allow Willstätter to continue with cocaine research as an independent investigator. This proved to be a blessing in disguise. With Baeyer's help, Willstätter started his work on tropine, a compound similar to cocaine. It was widely believed to have a similar structure as cocaine. By establishing the structure of tropine and by applying the data to the structure of cocaine, Willstätter showed that cocaine's structure was indeed different from the one proposed by Einhorn and coworkers. Although Willstätter was not the first to identify cocaine, he was the first to solve its structure and synthesize it in the laboratory. The immediate fallout of this was a complete breakdown of relationship with his Ph.D advisor. Einhorn refused to speak to Willstätter during the long years he spent at the University of Munich!

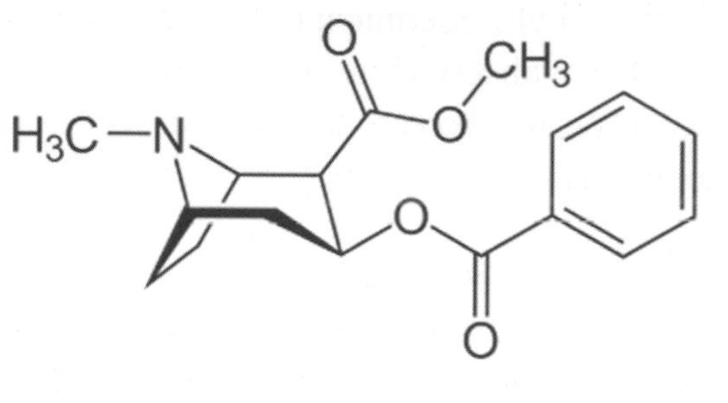

When Willstätter was studying at the University of Munich, Vin Mariani, a wine made from cocoa leaves was a popular drink among students in Europe. It acted as a stimulant and helped students to prepare for exams as it kept them awake and alert. Its secret ingredient was cocaine. We often consume it in many drugs especially cough syrups containing codeine, a derivative of cocaine.

It is a matter of coincidence that Willstätter solved the structure of cocaine first, and later of chlorophyll – two chemicals that have impacted our lives but with contrasting results.

Willstätter married Sophie Leser in 1903. Ludwig their son was born in1904 and their daughter Margarete, in 1906.

Chlorophyll

During his stay at the University of Munich, Willstätter worked on determining the structure of plant alkaloids like atropine and cocaine and synthesizing them in his laboratory. One of his principal objectives of this work as well as of his study of quinone and quinone type compounds was to develop the essential skills required to carry out studies based on chemical methods on dyes. This training stood him in good stead when Willstätter started research on plant and animal pigments. Unfortunately, the research facilities at the University of Munich were inadequate. Just as he was wondering how to solve this problem, he was offered a position as professor of chemistry in 1905 at ETH in Zurich.

Willstätter's move to ETH in Zurich was a turning point in his career. Even while he was at Munich, Willstätter had been intrigued by the complexity of chlorophyll and other pigments and their crucial role in plant and animal life. He decided to focus his research on understanding the structure and chemistry of the chlorophyll molecule. To do this, Willstätter had to first isolate pure and undamaged chlorophyll molecules present in leaves. Willstätter and his students succeeded in developing newer methods for extracting pure chlorophyll. After years of painstaking work using improved chromatographic techniques Willstätter and his students were able to collect extensive samples of chlorophyll from different plants. On comparing the samples they discovered that chlorophyll from different plants was essentially the same.

Willstätter returned to Germany after seven years in Zurich to take up a dual appointment of an honorary professorship at the University of Berlin and director of the Kaiser Wilhelm Institute of Chemistry with research facilities in Berlin/Dahlem. His institute was located next to the Institute for Physical Chemistry and Electrochemistry headed by Fritz Haber. The two became close friends.

The Kaiser Wilhelm Institutes in various science disciplines were established in 1910/1911 to promote natural sciences. Their main aim was to provide outstanding scientists opportunities to conduct research on problems of their own choosing without time consuming teaching responsibilities. After the Second World War, The Kaiser Wilhelm Institutes were renamed Max Planck Institutes.

Willstätter continued his work on chlorophyll even after his return and proved conclusively that chlorophyll was a combination of blue-green chlorophyll *a* and yellow-green chlorophyll *b* (in 3:1 ratio). Also, contrary to the assumption of earlier researchers working on chlorophyll that magnesium in the chlorophyll molecule was an impurity, Willstätter's study showed that magnesium was an essential component of the chlorophyll molecule. This discovery established for the first time the importance of magnesium as a plant micronutrient.

Chlorophyll and hemoglobin

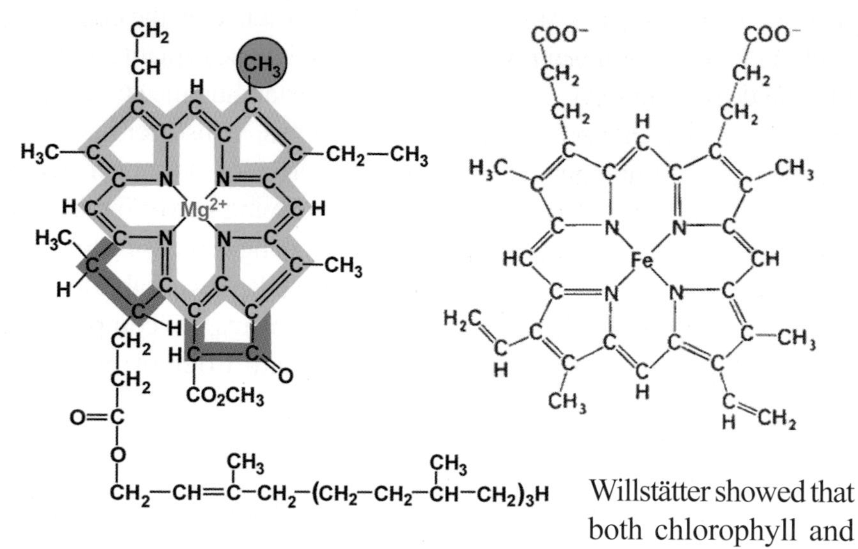

Willstätter showed that both chlorophyll and hemoglobin had a single atom at the centre surrounded by a ring-like structure but with an essential difference - while the central atom in chlorophyll was magnesium, the central atom in the hemoglobin molecule was iron. He also showed that heme, the component that gave blood its red colour, had a structure similar to porphyrin in chlorophyll.

Colours of flowers

Willstätter's other major contribution to chemistry was his monumental work on the chemistry of non-green plant pigments. His work to isolate chlorophyll from plant materials yielded an unexpected but important by-product – a yellow solution. A study of this solution revealed that it contained carotenoid pigments which gave carrots, egg yolk and tomatoes their unique colours.

In Zurich, apart from his study of chlorophyll, Willstätter had started a study of a group of chemical compounds (anthocyanins) to understand their role in red and blue colours of flowers. When he began his study, it was winter in Zurich and he could get only dried

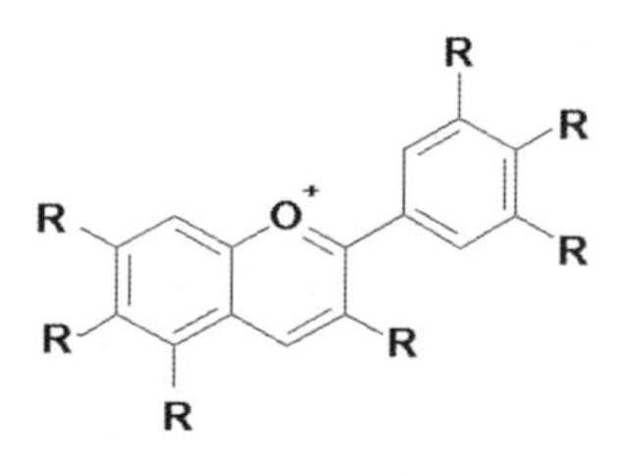

cornflowers. It was not a good choice as the dried flowers retained only a small amount of the pigment (less than 1%). When he joined the Kaiser Wilhelm Institute in Berlin, Willstätter raised fields of brightly coloured flowers like cornflowers, asters, chrysanthemums, pansies, and dahlias on the Institute grounds and in the garden around his residence so that he could have an unending supply of different coloured pigments for his research on anthocyanins. Between 1912 and 1916, he investigated anthocyanins and published 18 papers on his findings and demonstrated that anthocyanins and carotene present in plant leaves were mainly responsible for the colours of autumn foliage. He was able to show that colours of fruits such as plums, cherries and berries were mainly due to chlorides of pelargonidin, cyanidin and delphinidin - three closely related compounds.

The outbreak of World War brought further research to an abrupt end as his co-workers were drafted into military service and the army took over his Institute. The flowers from his rather neglected garden went to military hospital to cheer the wounded soldiers rather than to his laboratory for research. This interruption affected Willstätter deeply and he did not return to research on colours of flowers and fruits even after the war ended!

War efforts

During the war, his friend Haber in his capacity as the person in charge of chemical warfare work was developing a poison gas to be used against the enemy soldiers. He requested Willstätter to assist him in this war effort. Willstätter refused to work on poison gas but agreed to work on developing a gas mask that could absorb chlorine and phosgene gases. He and his coworkers developed in a record time of five weeks a gas absorbing unit consisting of a canister of activated charcoal and urotropin or hexamethylenetetramine that absorbed all of chlorine and phosgene. Willstätter was awarded the Iron Cross (Second Class) by the German government for this work (From Nobelprize.org [2]). It was during his work on making a good gas mask that he learnt he had been awarded the 1915 Chemistry Nobel Prize. A year later, he was offered the prestigious professorship at Munich University after Baeyer's retirement. He was also offered the position held by Otto Wallach (who had just retired) at Gottingen. Though Willstätter was tempted by this offer as Gottingen being a medium sized university offered greater scope for interaction among colleagues in different departments, he accepted the offer from Munich since his demands for increased lab space and creation of a full professorship in physical chemistry were accepted by the authorities.

Fritz Haber was awarded the Nobel prize in 1918 for ammonia synthesis. What is unbelievable about Haber is that a person who saved humanity through the synthesis of ammonia also made poison gas to kill innocent human beings.

Research at Munich

In Munich, Willstätter began a detailed study to understand the nature of enzymes and how they functioned. His efforts to develop better and newer methods to isolate and purify enzymes and his foray into photosynthesis are regarded as precursors of biochemistry of today. His claim that enzymes were not proteins however proved to be wrong.

Willstätter worked with Friedrich Bergius on the hydrolysis of cellulose with hydrochloric acid to obtain dextrose. Dextrose was then fermented to get alcohol. This process to obtain alcohol was refined by others later but came to be known as the Bergius-Willstätter process.

One of the most outstanding achievements of Willstätter was the preparation and prediction of the correct structure of cyclooctatetraene (C_8H_8). He and Heidelberger, his post doctoral student from America, succeeded in perfecting the synthesis of this hitherto elusive compound. (Heidelberger went on to become one of the greatest immunologists). Willstätter achieved this by sheer chemical intuition. (How did he do this in the absence of any physical methods?). The discovery of cyclooctatetraene (C_8H_8) by Willstätter in 1911 shattered Thieles' partial valence theory of aromaticity which had proposed that all fully conjugated cyclic polyenes should display benzene-like aromatic character.

Again, the conflict between personal happiness and professional satisfaction surfaced as he had to confront more pronounced overt anti-semitism at the university. While he was able to deal with it at a personal level, it impacted on his relations with others at the place of work.

Honours

Willstätter was awarded the Nobel Prize for Chemistry in 1915 *"for his researches on plant pigments, especially chlorophyll"* (From Nobel Lectures [3]). Willstätter learnt of this news while he was busy with his work on finding the right solution for the absorbing unit of the gas mask. Interestingly he could receive the prize in 1920 only after the World War ended in 1919. Max Planck, Fritz Haber, Johannes Stark and Max von Laue, other German scientists awarded the Nobel Prize during first World War years also received it in a special ceremony held in Stockholm in 1920!

Willstätter received many other honours and awards for his outstanding contributions to chemistry. Some of them are: Davy medal of the Royal Society, London (1932), the Willard Gibbs Medal of the Chicago Section of the American Chemical Society (1933) and foreign membership of the Royal Society, London (1933).

Willstätter the man

Willstätter was a man of great integrity and loyalty. He stood by his younger colleagues and tried his best to protect them from the fallout of both covert and overt anti-semitism in Germany during the early 1900s. While he handled the personal problems resulting from this scourge with forbearance and dignity, never letting it affect his research, he was unwilling to accept it when it impacted on his authority as the director of the research laboratory. When the appointment of Professor Victor Goldschmidt of Norway was turned down by the faculty solely on the basis of his race, Willstätter resigned from his job. His research at the institute came to an abrupt end as he refused to reconsider his decision in spite of overwhelming requests from his colleagues. He left his job in 1925.

"Expressions of confidence by the Faculty, by his students and by the Minister failed to shake the fifty-three year old scientist in his decision to resign. He lived on in retirement in Munich....Dazzling offers both at home and abroad were alike rejected by him" (From Nobel Lectures [3]).

Three years later, Heinrich Wieland who succeeded Willstätter accommodated Willstätter's assistant Margarete Rohdewald in Willstätter's private laboratory to do collaborative research with Willstätter. For nine years starting from 1929, Willstätter and Margarete Rohdewald did research on enzymes and published 18 papers. It was an unusual collaborative work as it was conducted almost entirely through telephonic conversations. Willstätter who was a meticulous experimentalist did not see her working in the laboratory during this period!

Willstätter was a generous teacher and a research guide and made lifelong friends. This aspect of his personality is best illustrated by the experience of Michael Heidelberger, a young and not so well to do American chemist who came to work with him in 1911. They came to an agreement about how to bear the cost of laboratory supplies. It was a simple arrangement-whenever expensive items (for example silver nitrate needed for research) were bought, Willstätter would pay and Michael Heidelberger would pay for the inexpensive materials such as sulphuric acid! At the end of his stay when asked about his experience of working with Willstätter, Heidelberger seems to have remarked "*Better training than that you couldn't have*" (From "Richard Willstätter." Wikipedia [4]). Willstätter and Heidelberger remained friends for life.

Willstätter's personal life was marked by tragedies. First, his wife Sophie Leser whom he married in the summer of 1903 died tragically after just five years leaving behind a four year old son, Ludwig, and two year old daughter, Margarete. Sophie died due to peritonitis caused by the delay in surgical intervention after her appendix had ruptured. He was devastated by this and as he remarked in his autobiography he immersed himself in his work and taking care of his two young children and did not take a vacation for ten years. In the spring of 1915, ten-year old Ludwig died without warning from suspected juvenile diabetes. Willstätter tried to come to terms with this blow by immersing himself in his work on developing a gas mask. It seemed that work was the antidote for his personal tragedies. *"His panacea...... of all the vacillations of fortune was 'work' and then 'more work'"* (From Robinson [5]).

Last days

In 1938, the oppression and hounding from the Gestapo became unbearable. His former student and colleague A. Stoll helped Willstätter escape to Switzerland leaving behind most of his belongings. He settled down at Muralto near Locarno. Willstätter spent the remaining years of his life writing his autobiography *Aus meinem Leben* in which he gave *"a clear picture of his happiness and exaltation as well as grievous disappointments and tragic losses"* (From Robinson [5]). He died in Switzerland on August 3, 1942 of cardiac arrest at the age of 70 years.

References

1. Willstätter, Richard, *From My Life, The Memoirs of Richard Willstätter*. Translated by Lilli S. Hornig. New York: W.A. Benjamin, 1965.
2. The Nobel Prize in Chemistry 1915, *Nobelprize.org*. Nobel Media AB 2014. Web. 21 Apr 2015. http://www.nobelprize.org/nobel_prizes/chemistry/laureates/1915/
3. From *Nobel Lectures, Chemistry 1901-1921*, Elsevier Publishing Company, Amsterdam, 1966.
4. Wikipedia contributors. "Richard Willstätter." *Wikipedia, The Free Encyclopedia*. Wikipedia, The Free Encyclopedia, 4 Dec. 2014. Web. 21 Apr. 2015.
5. Robert Robinson, *Willstätter memorial lecture,* J. Chem. Soc.,1953, 999-1026.

"Richard Willstätter - Biographical". *Nobelprize.org*. Nobel Media AB 2014. Web. 21 Apr 2015. *<http://www.nobelprize.org/nobel_prizes/chemistry/laureates/1915/willstatter-bio.html>*

15. GILBERT NEWTON LEWIS (1875–1946)

The 20th century chemical genius who did not get the Prize

"The half century which terminated with the death of Gilbert Newton Lewis will always be regarded as one of the most brilliant in the history of scientific discovery, and his name ranks among the highest in the roster of those that made it great. The electron theory of chemical valence, the advance of chemical thermodynamics, the separation of isotopes which made possible the use of the deuteron in the artificial transmutation of the elements, the unraveling of the complex phenomena of adsorption, fluorescence, and phosphorescence of the organic dyes are among the achievements which will ever be associated with his name" (From Gettell et al. [1]).

(From Edgar Fahs Smith Collection)

G. N. Lewis is one of the greatest physical chemists of all time. He summed up the subject as *"encompassing everything that is interesting"* (From Helrich [2]). He is remembered as one of the pioneers who brought education and research in physical chemistry and chemical physics to the forefront in the US. In the 19th century and early 20th century, European scientists, particularly physicists, dominated the scientific scene. Gilbert Newton Lewis by his outstanding research in a number of interdisciplinary fields and by his innovative approach to chemical education, led a revolution that ended the European dominance in physical chemistry. Lewis was one of the greatest chemists of the 20th century with an immense impact on chemistry as a whole.

Early years

Lewis was born to Frank Wesley Lewis, a man of strong and independent views and a lawyer by profession, and Mary Burr in 1875 at Weymouth, Massachusetts, USA. As Gilbert Lewis was a precocious child and could read well when he was only three years old, his parents decided to educate him at home instead of sending him to school. This decision appears to be a forerunner of his unorthodox approach to scientific problems and his ability to think out of the box. He started formal schooling in 1889 only at the age of 14, five years after his parents moved to Lincoln, Nebraska, by joining the preparatory school of the University of Nebraska there.

After successfully completing the preparatory school, Lewis joined the university as an undergraduate in the department of chemistry and moved to Harvard University after two years. He received his B. S. degree in 1896 from Harvard. He then took a year off and taught at the Philips Academy in Andover, Massachusetts. He rejoined Harvard University in 1898 to work under T. W. Richards and completed both his Master's and doctoral work in just two years. After obtaining the Ph.D degree in 1899, Lewis received a travelling fellowship to do postdoctoral work. He decided to go to Germany for the purpose and spent a semester each working with Ostwald at Leipzig and Nernst at Gottingen. It is widely believed that Lewis had academic differences with Nernst and Nernst did not seem to have forgiven Lewis for what he considered as Lewis' audacity to raise questions about his work.

Early academic career

After a year in Europe, Lewis returned to Harvard in 1901 as an instructor in electrochemistry and thermodynamics. After three years in the job, Lewis got a year off. He had the feeling that his unusual abilities were not recognized by the conservative senior faculty at Harvard. His shyness and perhaps lack of self-confidence at that time prevented him from asserting his views. He did not either pursue or publish some of his ideas (such as his cubic atom and his speculation about light having pressure). Both these ideas later formed the basis of his theory of the chemical bond as formed by the sharing of electrons (1916). While lack of academic freedom irked Lewis enough to decide to leave Harvard, many of his research ideas related to thermodynamics, relativity and valence were kindled during his stay at Harvard. He decided to do something totally different from what he was doing at that time.

He accepted the position as superintendent of the weights and measures in the Bureau of Science at Manila, Philippines and left the U S in 1904. He took this unusual break so that (as he said later) he could think for himself his course of future action. He took Nernst's book "Physikalische chemie" with him and corrected all the errors. He developed his love for good cigars during his stay in Manila. Even at Manila, Lewis found time from his formal job to pursue experimental research on some interesting problems.

Lewis returned to the U S a year later in 1905 when he accepted the offer of a faculty position at the Massachusetts Institute of Technology (MIT). He joined the physical chemistry group working with Professor A. A. Noyes. This group of physical chemists gave a strong impetus to the growth of physical chemistry in the country. Lewis stayed at MIT for seven years (1905–1912) and these years laid the foundation for his lifelong interest in thermodynamics. It was at MIT that Lewis started his path-breaking work on measuring free energies. At that time, chemists could not apply principles of thermodynamics to understand chemical equilibrium in reactions. Lewis found a method to measure free energy of formation of compounds. This period also saw Lewis maturing as a scientist who could simultaneously carry out studies on both theoretical and experimental problems. He was a polymath and could concentrate on a number of problems at the same time. His prodigious scientific output of this period include determining accurately the electrode potentials of a number of elements (Lewis was proud of this work), publishing two seminal papers- "Outlines of a New System of Thermodynamic Chemistry" and "Thermodynamics and the

Free Energy of Chemical Substances." He became a full professor at MIT in 1911 when he was 36 years old.

Lewis considered himself to be both a chemist and a physicist. He was interested in many problems in physics even when he was a young researcher. He carried out studies on radiation, particularly the relation between mass and energy. After meeting Einstein, post publication of the theory of relativity in 1905 and of mass-energy, Lewis decided to revisit his early work where he had deduced his equation (for mass-energy) using the pressure of light. His work had similarities with Einstein's famous equation and he became one of the early supporters of Einstein's theory of relativity. Lewis took great delight in the complexities in Einstein's theory.

Lewis became known for his unorthodox and innovative approach to scientific problems. He had also accomplished results in thermodynamics which had caught the attention of the scientific community. In 1912, when he was 37 years old, University of California, Berkeley offered Lewis the dual positions of Chairman of the Department of Chemistry and Dean of the College of Chemistry. Even though UC Berkeley was not as well known as MIT at that time, and its chemistry department was in a bad shape and in need of restructuring its research and teaching activities, Lewis decided to accept the offer as he was guaranteed complete freedom in his choice of personnel and implementation of his ideas in building a department of his dreams. He joined UC, Berkeley in 1912 and stayed there till he passed away in 1946. His greatest achievement which rivals

(From Wikimedia commons)

South Hall, University of California, Berkeley

even his monumental contributions to a number of areas in chemistry is the unique department of chemistry he built and nurtured.

Lewis and Thermodynamics

Lewis had worked in thermodynamics in 1899 for his doctoral work at Harvard. Based on this work, he published his first paper on thermodynamics (jointly with T.W. Richards, his research supervisor) "Some Electrochemical and Thermochemical Relations of Zinc and Cadmium Amalgams". Thermodynamics at that time was considered a domain of physicists, and chemists found it difficult to apply it to solve chemical problems. Even well-established laws of thermodynamics (including the law of conservation of energy) were not systematized and chemists were aware of them as stand alone equations rather than as a system that logically led to determining the chemical equilibrium in a reaction. Lewis showed that the equilibrium data on chemical reactions could not be obtained by measuring electrode potentials. He began working both on theory and experiments to correlate the methods of physics and chemistry towards understanding thermodynamics and published a seminal paper titled "Outlines of a New System of Thermodynamic Chemistry".

> Chemists had desired for a long time to have a comprehensive table giving information of chemical affinity to aid them in anticipating the direction of a chemical reaction. It was believed at that time that the heat of reaction could indicate chemical affinity and as a result, heat of formation of substances participating in the chemical reaction could be used to determine the direction in which the reaction would proceed. The emergence of thermodynamics changed this belief and showed that free energy and not heat gave the correct measure of chemical affinity. Lewis prepared a reliable and comprehensive table containing free energy data. To accomplish this task, Lewis and his co-workers examined chemical equilibria in many reactions.

Lewis was aware of the earlier work of Josiah Willard Gibbs who had shown that it was possible to determine chemical equilibrium of a reaction on the basis of the free energies of the reacting substances. There were also the work of Berthelot (France) and Julius Thomsen (Denmark) on the enthalpies of chemical substances as well as the experimentally derived laws on the behaviour of dilute solutions and ideal gases proposed by Ostwald (Germany), van't Hoff (Netherlands) and Arrhenius (Sweden). Lewis took up the challenge of reconciling the apparently conflicting theories and of finding a method for making established empirical laws applicable to real systems. He realized that to succeed in this effort, he had to either measure the free energy of chemical systems or calculate entropies via enthalpy values which were already known. He conducted a number of experiments to arrive at accurate values for free energy and entropy. Throughout his work on chemical thermodynamics, he introduced new terms and concepts such as fugacity in 1901 (while at Harvard), activity coefficient in 1907 (while at MIT) and ionic strength in 1921 (while at Berkeley).

Young Lewis introduced the term fugacity or 'escaping tendency' of a real gas (not an ideal gas). Fugacity was the tendency of a chemical substance (particularly gaseous substance) to move spontaneously from one chemical phase to another. He was convinced that the principle of fugacity was necessary to arrive at thermodynamic relations. He derived a formula to calculate fugacity and devised an experiment to determine its value. The concept of fugacity survives and is still used to describe behaviour of real gases.

Lewis measured free energies of compounds of oxygen, nitrogen, halogens, sulfur and alkali metals and several simple organic compounds. In 1912, he measured the free energies of the formation of urea and ammonium cyanate. His insight on the importance of the role of free energy in complex organic systems proved prophetic. In 1912, Lewis expounded a new approach to the understanding of thermodynamics based on the concept of activity, a term he introduced. According to him, activity was a *"function with the dimensions of concentration"* and indicated the *"tendency of substances to cause change in chemical systems"* (From Encyclopedia.com [3]). Even though the concept of activity (much like his idea of fugacity) may not have played a crucial role in the theoretical understanding as Lewis believed it would, it has proved indispensable for measuring deviations in real solutions from ideal behavior. Realising the utilitarian value of a complete table of free energies for predicting chemical behaviour, he was convinced that data collection for creating such a table was "*an imperative duty of chemistry*" (From Encyclopedia.com [3]). From 1913 to 1920, Lewis worked tirelessly in this effort.

Lewis published a series of important papers (many of them with Merle Randall) on thermodynamics - the two most important papers among them being one giving entropy data and the other giving the results of the empirical verification of Nernst 's heat theorem. Lewis actually modified Nernst's heat theorem to derive the now accepted third law of thermodynamics which takes entropy into account. He also carried out experiments to verify the law experimentally by low-temperature measurements.

Lewis occupies an important position in the development of thermodynamics for devising methods for practical applications of the laws to real chemical systems. His outstanding contribution was in simplifying its laws so that they were relevant to real systems. Even at the nascent stage of the subject, he was regarded as next only to Nernst and Haber in importance in the field.

Lewis published his monumental treatise "Thermodynamics and the Free Energy of Chemical Substances" with Merle Randall in 1923. This soon became the most popular text book of chemical thermodynamics used by generations of chemists as it presented clearly what were considered difficult concepts and contained extensive and reliable data of free energies. Most importantly, chemical thermodynamics was no longer the privilege of experts and its tools came to be accepted by all chemists as essential for both doing research and for chemical education. The preface of the book itself is a masterpiece of chemical literature.

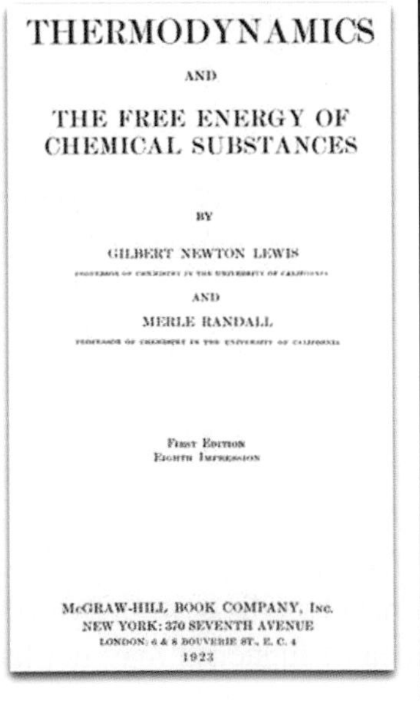

THERMODYNAMICS

AND

THE FREE ENERGY OF
CHEMICAL SUBSTANCES

BY

GILBERT NEWTON LEWIS

AND

MERLE RANDALL

FIRST EDITION
EIGHTH IMPRESSION

McGRAW-HILL BOOK COMPANY, INC.
NEW YORK: 370 SEVENTH AVENUE
LONDON: 6 & 8 BOUVERIE ST., E. C. 4
1923

Theory of Valence and the Chemical bond

The beginning of the 20th century witnessed great interest among physicists and chemists to understand the structure of the atom. Chemists sought to investigate the role of valence of an element and how electrons in atoms formed chemical bonds. Richard Abegg of Germany recognized early that the arrangement of eight electrons (which was reached when an atom lost or gained the required number of electrons and thus became either a positive or a negative ion) was crucial to attain stability by atoms. Lewis was the first chemist to provide a model for this arrangement. While giving a lecture on the laws of valence at Harvard in 1902, Lewis wanted to make the law easier to understand by providing a model. It occurred to him to represent an atom as a "cubic atom" – i.e. representing an atom as made up of concentric cubes having eight corners with each corner representing possible positions of the outermost electron. Lewis used the

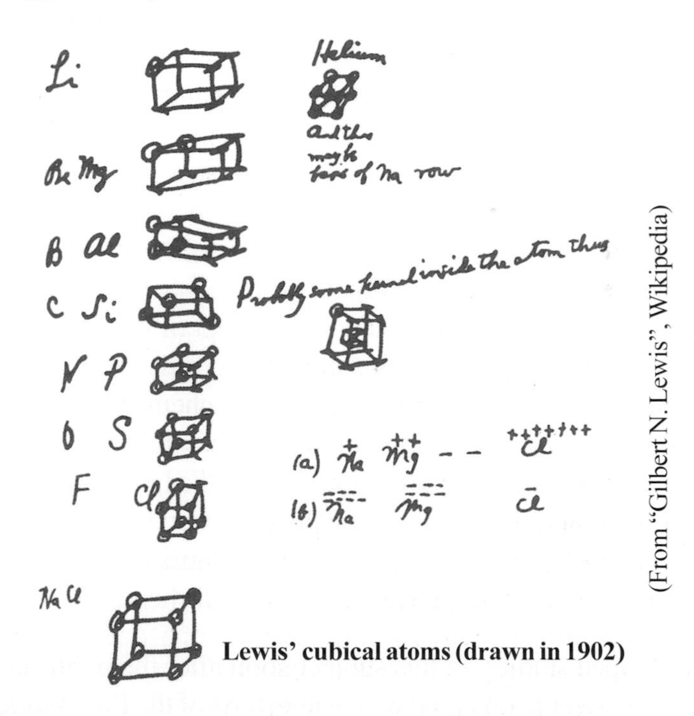

Lewis' cubical atoms (drawn in 1902)

(From "Gilbert N. Lewis", Wikipedia)

original drawings (as yet unpublished) of cubical atoms in his lecture notes. As he considered them as the first attempt to provide a concrete model, he referred to them in his classic 1916 paper on the chemical bond. Historically, this is the first proper paper on the chemical bond. His model corroborated the prevalent idea that each atom acquired eight electrons in its outer shell. Unfortunately, his explanation of how chemical bonds were formed was not accepted by his superiors at Harvard and he felt so discouraged that he did not publish his theory of the cubic atom until he moved to Berkeley. According to Lewis, for a chemical bond to be formed, there should be four pairs of electrons surrounding each atom. Using this model, Lewis derived the structures of oxy acids, ammonium ion and of halogen molecules (which was not possible to derive using the valence theories prevalent at that time).

Lewis was the first to realize the crucial role played by the valence electrons in the outer most shell in forming chemical bonds. Several chemists started working on his theory of octet formation and the theory of how chemical bonds were formed continued to evolve. In his 1916 paper "The Atom and the Molecule", Lewis collated data from his earlier work on valence and chemical bonds and defined a chemical bond as a bond formed by a pair of electrons jointly held or shared by the two atoms. At first he illustrated this concept of a single bond by two cubic atoms sharing a corner. He simplified this further by placing two dots between two atoms. He introduced this diagram to represent the electronic structure of atoms and molecules. These electron dot diagrams came to be known as Lewis dot structures. He also introduced the concept of the covalent bond – a bond which is formed not by transfer of electrons but by sharing them. This model found acceptance of both organic chemists and co-ordination chemists. Sharing electron pairs could be successfully used to understand and explain complex organic reactions. Use of the Lewis' model in organic chemistry was promoted enthusiastically by Robert Robinson, Christopher Ingold and Arthur Lapworth in England. Maurice Huggins and Nevil Sidgwick promoted this model in coordination chemistry.

Lewis stopped studies on this subject soon after the publication of this paper as he became involved with war efforts of the First World War. He was also disappointed that his shared electron theory had no takers in 1916.

> Irving Langmuir started working on the Lewis' theory in 1919, Langmuir was a skillful communicator and his lectures and papers on the new theory of the structure of the atom brought Lewis' theory into focus. He was so successful that Lewis' theory soon came to be known as Lewis-Langmuir theory or simply as Langmuir theory. Lewis was upset by the credit given by chemists to Langmuir for the original idea proposed by him. There was much bitterness in the relationship between Lewis and Langmuir. Eventually, Langmuir stopped working in this area and gave full credit to Lewis.

The work of Lewis on the relation between valence of an atom and chemical bonding reached its climax in 1923, when he published "Valence and the Structure of Atoms and Molecules" as a monograph of the American Chemical Society in which he brilliantly presented the highlights of his model. He was able to bridge the gap between the views of organic and inorganic chemists on bond formation by his concept of paired electrons. He also explained the nature of nonpolar bonds. This monograph had a profound impact on the way chemists thought about chemical bonding between atoms and molecules. After the publication of the monograph, Lewis worked only sporadically on the subject. Pauling using quantum mechanics transformed Lewis' seminal concept of shared electrons into the valence bond model and published the classic, "The Nature of the Chemical Bond" in 1939. Lewis richly deserved to get the Nobel Prize for expounding the most fundamental theory in chemistry but it was not to be.

VALENCE
and
The Structure of Atoms and Molecules

BY
GILBERT NEWTON LEWIS
PROFESSOR OF CHEMISTRY IN THE UNIVERSITY OF CALIFORNIA

American Chemical Society
Monograph Series

BOOK DEPARTMENT
The CHEMICAL CATALOG COMPANY, *Inc.*
19 EAST 24TH STREET, NEW YORK, U. S. A.
1923

Acid and Bases

Svante Arrhenius is credited with being the first to give a modern definition of acids and bases in 1884. He defined any element or compound that disassociated upon dissolving in water to give hydrogen ions or H^+ as an acid and a base as any molecule or compound that formed OH^- or hydroxyl ions on dissolving in water. According to this model, in an acid–base reaction, a hydrogen ion is removed from the acid and this gets added to the base resulting in the formation of a salt and a solvent. Lewis felt that this definition of an acid or a base was one-sided. Brønsted and Lowry defined acid-base reactions in terms of proton transfer giving rise to conjugate acids and bases in an acid base reaction. The main drawback of these models was that for any substance to be classified as an acid, presence of hydrogen was essential.

In 1938, Lewis stated *"To restrict the group of acids to those substances that contain hydrogen interferes as seriously with the systematic understanding of chemistry as would the restriction of the term oxidizing agent to substances containing oxygen"* (From Hall [4]). Lewis took a completely new approach to define acids and bases. While working on his shared electron theory as the basis for the formation of chemical bonds, Lewis noticed there were instances where the electron pairs were not shared by participating molecules. To explain this anomaly, he called such a molecule an odd molecule (now called free radical) which does not share the electron. Lewis proposed a newer definition of acids and bases on the basis of structure of the molecules and their bonding and proposed his electron pair theory of the reaction between an acid and a base. Accordingly, a Lewis acid is an ion or molecule which is an electrophile (electron loving) with vacant orbitals. In other words, the outer most shell of a Lewis acid has an incomplete octet. As a result, it accepts an electron pair or a pair of non-bonding valence electrons. He defined a base as *"..... one which has a lone pair of electrons which may be used to complete the stable group of another atom, and that an acid is one which can employ a lone pair from another molecule in completing the stable group of its own atoms. In other words, the basic substance furnishes a pair of electrons for a chemical bond and the acid substance accepts such a pair"* (From Lewis [5] & Hildebrand [6]). In other words, Lewis base is a compound which is a nucleophile with lone pair electrons and therefore donates an electron pair. This definition accomplished two things at the same time- the properties of a base was not necessarily the properties of a hydroxyl ion and the acid no longer had to be a proton donor. According to Lewis, any compound that can accept a pair of electrons is an acid and any compound that can donate a pair of electrons is a base. The acid–base reaction between a Lewis acid and a Lewis base does not result either in the formation of a salt and a solvent or in a conjugate acid and a conjugate base but results in a coordinate covalent bond forming an adduct (an addition compound) where the outermost shell of all atoms fulfill the octet rule. The best example of Lewis acid-base reaction is the one between BF_3 and NH_3. As Lewis acids are defined in terms of structure, it gave flexibility as chemists could now view all molecules and compounds that accept electron pairs as acids. Another advantage of this model over the other two models is that it covers oxidation-reduction as these reactions also involve transfer of electrons.

Heavy water

After the long and intense period of scientific activity stretching over two decades, there was a lull after the publication of his acid-base theory. Lewis was bored of working on collecting free energy data as well. For the next decade, starting from 1923, he worked on quantum radiation and tried to formulate the laws of quantum radiation using tools of thermodynamics. Though he was encouraged by a letter from Einstein, he soon realized that he could not arrive at any satisfactory theory or explanation. However, the term *photon* that he coined got accepted and is used even today. His restless mind was looking for fresh challenges.

The discovery of deuterium by his former student Harold Urey in 1932 opened up a new field for Lewis – isotope separation. Lewis was working on making oxygen isotopes at that time. Urey's suggestion that electrolysis could be used to isolate deuterium on a large scale, made Lewis immediately realize that it would be easier to isolate deuterium, an isotope of hydrogen, as it was twice as heavy as hydrogen. Lewis also realized that compounds of deuterium would be quite different from the compounds containing hydrogen and would have unusual properties. He succeeded in making pure D_2O (heavy water) in 1933. Between 1933 and 1935, Lewis published 28 papers on various aspects of the chemistry of heavy water. He conducted experiments to determine the effect of heavy water on germinating seeds as also its effect on mice fed on D_2O. He collaborated with E. O. Lawrence to study the behavior of heavy water in a cyclotron. Lewis again opened up an entirely new field (heavy water) and yet again missed the Nobel Prize!

(From www.pbs.org)

Gilbert Lewis isolated the first sample of pure heavy water from ordinary water in 1933.

Photochemistry

Having been disappointed by the lack of proper recognition for his work on heavy water, the restless mind of Lewis found another unexplored area to conduct research which combined theoretical studies with experimental work. He initiated studies to understand the relation between colours in compounds and photochemistry. In photochemistry, Lewis discovered a field that challenged his inquiring mind and experimental expertise. He conducted

a number of experiments with Melvin Calvin on phosphorescence and fluorescence. Lewis published a review on their theory of colour of compounds in 1939 and this was followed by a series of papers. Lewis proposed that the transition of a molecule from its triplet state to a lower energy singlet state caused phosphorescence. The major papers he published between 1944 and 1945 with Calvin and later with Michael Kasha, his last student, announcing this important conclusion marks a new direction in photochemistry. This concept was not accepted by Edward Teller and James Frank among others as quantum mechanics rules did not allow such a transition. However, the original idea of Lewis proved correct later.

Honours

Lewis was honoured with honoris causa degrees by many universities, elected to memberships of science academies and conferred many prestigious awards and medals. A few of them are:

Honoris Causa from the universities of Liverpool (UK), Madrid (Spain), Chicago, Wisconsin and Pennsylvania (USA).

He was elected as a foreign member of the Royal Society of London, the Chemical Society of London, the Royal Institution of Great Britain, the Indian Academy of Sciences, the Danish Academy and the Swedish Academy.

He was awarded the Nichols medal (1921), Gibbs medal (1924), Davy medal of the Royal Society (1929), Richards medal (1938) and Arrhenius medal (1939) of the Royal Swedish Academy of Sciences.

In 1925 he was selected to deliver the prestigious Silliman Lecture at Yale. The title of his lecture was "The Anatomy of Science". A year after he moved to University of California, Lewis was elected to the National Academy of Sciences in 1913 (one of the youngest to be elected) and from which he resigned in 1934. It was also in 1934, he along with Pauling and Noyes founded the *Journal of Chemical Physics*.

Chemistry at Berkeley

Just as many chemists regarded the greatest discovery of Davy to be Faraday, many regard the chemistry department Lewis built and nurtured at UC Berkeley to be his greatest contribution. He set the trend and established traditions both in teaching and research that have survived more than a century making the department chemistry at UC, Berkeley numero uno chemistry department in the world even today. When Lewis left the well-established group under Noyes at MIT and joined UC Berkeley in 1912, the chemistry department was not in a good shape in the university that was just trying to establish itself. How did Lewis, then only 37 years old, approach the stupendous task and succeed in it without sacrificing his own research? His first action was to persuade his colleagues Hildebrand, Tolman and Bray to go west with him in this exciting adventure. This proved to be a masterstroke as all of them were brilliant and committed teachers and wrote books which became classics. He used the freedom he had been given to recruit a core group of enthusiastic young scientists who were excited by the task of establishing a new department of chemistry. Lewis' vision of a chemistry department was one where there was no artificial compartmentalization of different divisions of chemistry. Weekly group meetings were held where members discussed problems related to all of chemistry and every member enjoyed complete freedom to express his/her opinion. Lewis took great joy in deliberately expressing contradictory views to shake up prejudices and preconceived notions and initiate lively discussions. Every new idea was whetted by frank discussion. As a result according to Hildebrand *"The whole department thus became far greater than the sum of its individual members"* (From Hildebrand [6]). Another great input from Lewis that contributed to the cohesiveness of the department was his unfailing defense of his colleagues and making sure that they were not overloaded with excessive contact hours. There was complete academic freedom and truly democratic traditions were established.

Lewis believed in a vibrant curriculum and decided to radically change the approach to curriculum right from the introductory classes. The number of formal lectures were limited so that even undergraduate students could undertake research problems. The curriculum was broad based to avoid narrow specialization. There were weekly research discussion meetings where a wide variety of research topics – even speculative ideas - were discussed freely with both faculty and students – an experience he had missed in his student days at Harvard. The undergraduate curriculum in chemistry consisted of a few essential basic courses and the students were offered a great choice of electives in their senior years. Hildebrand even formulated this empirical academic law *"the number of courses taught by a department varies inversely with the eminence of its faculty and with the advancement of knowledge in the field"* (From Hildebrand [6]). Lewis modernized chemical education which became the template for other American universities to follow.

According to Hildebrand, "*one of Lewis' first moves was to turn almost the entire staff loose upon the problem of guiding the freshman in the way he should go, by fostering a scientific habit of mind. We met weekly to discuss the organization of the freshman course and the methods of presenting difficult topics. Although the lectures were given to five hundred students at a time in a huge chemistry auditorium with great attention to lighting, projection and realization of the full dramatic possibilities of the subject, the laboratory work and quizzing took place in sections of only twenty-five students, taught by a majority of the permanent staff with the help of numerous teaching assistants. The complaint that a freshman in a large university has no contact with professors did not apply in freshman chemistry at the University of California, Berkeley, for as many as eight full professors taught freshman sections in a single term. The example thus set by senior professors has had a profound effect upon the apprentice teachers, making them take their teaching seriously and convincing them that talent for research is not demonstrated by indifferent teaching*" (From Hildebrand [6]).

At the graduate level, research took precedence over lectures. Lewis was of the opinion that giving lectures on subjects/topics already available in textbooks was a waste of time. Students were encouraged to conduct research on topics of their choice. Attendance at the weekly seminars where students and faculty presented their work was mandatory for all the teaching staff and students. Every Tuesday the entire chemistry department headed towards Room 102, Gilman Hall where the seminar was held. The seminar began precisely at 4 pm and late comers were not allowed to enter the hall. Though Lewis did not formally teach either undergraduates or graduate students, he presided over these research conferences held every week and sat through every presentation in his favourite chair with his trademark cigar.

Lewis achieved this not by coercion or by invoking his authority, but by his unique ability to reason patiently with those who held different views and persuade people to arrive at a consensus. He held genuine consultations with all stake holders of the department before he took important decisions. This made the members of the department fiercely loyal and committed to the department of chemistry. This model is clearly the best template to build a successful department in a university.

> At a meeting of the faculty convened immediately after the Pearl Harbour attack by the Japanese to discuss its implications, someone in the room asked Lewis what he would do now for his cigars from Philippines. Lewis after being silent for a few minutes apparently said "the problem is more serious than I thought." He then ordered his secretary to buy all the stock of the cigars from his supplier!

Lewis the man

Lewis was a man of wide interests and he was a voracious reader. He was particularly fond of reading detective stories and books on history. In fact, he said *"A detective with his murder mystery, a chemist seeking the structure of a new compound, use little of the formal and logical modes of reasoning. Through a series of intuitions, surmises, fancies, they stumble upon the right explanation, and have a knack of seizing it when it once comes within reach"* (From Lewis [7]). Lewis changed his research fields effortlessly throughout his career. In fact as Hildebrand remarked *"few men in their sixties have the imagination to branch out in new directions. Most of his (Lewis's) work on photochemistry and heavy water was done after he was 65 years old"* (From Hildebrand [6]). Even though he was shy and almost a social recluse, in the company of his friends or in the laboratory he was in his elements regaling his friends with his wonderful sense of humour, jokes and funny anecdotes. Lewis loved good life especially his cigar.

According to Hildebrand he hated humbug and was intolerant of pretence. Respecting Lewis' reluctance to public display of affection even from his students and close colleagues, parties held to felicitate his achievements were small and intimate with no long formal speeches. Unlike many other scientists, Lewis almost never left Berkeley to meet other important and influential contemporary scientists both in US and Europe. He considered himself an iconoclast and deliberately expressed unorthodox and contradictory views as if to provoke a discussion.

He resigned from the membership of the US National Academy of Sciences since he did not approve of the way it scrutinized the candidates for election. (The only other person who resigned for the same reason was Richard Feynman). He expressed his scientific findings boldly without caring for the reputation of those whose beliefs he challenged. Lewis perhaps unknowingly paid a heavy price for this. While a number of his students (Seaborg, Urey, Calvin) got the Nobel Prize, Lewis himself was denied the honour even though the Nobel Prize archives reveal that he was nominated a record number of times (35) starting with his contributions to chemical thermodynamics. Each time he was not awarded the Prize in one field, he picked up another unexplored field and made a fundamental contribution to it. The contributions of Lewis to chemical bond were ignored as also his discovery of a new class of acids and bases. His work on photochemistry was not recognised and his studies of heavy water seem to have made no impression. Clearly, Lewis was the greatest genius of the 20th century who was denied the Nobel Prize. It is believed that Nernst and Arrhenius never excused him for having modified and expanded their work!

Lewis may have been a poor public speaker but he was an outstanding writer. Some of his writings are like great literary prose. The preface to his book on Thermodynamics is a classic example of this. Lewis is almost poetical when he wrote *"There are ancient cathedrals which, apart from their consecrated purpose, inspire solemnity and awe. Even the curious visitor speaks of serious things, with hushed voice, and as each whisper reverberates through the vaulted nave, the returning echo seems to bear a message of mystery. The labor of generations of architects and artisans has been forgotten, the scaffolding erected for their toil has long since been removed, their mistakes have been erased, or have become hidden by the dust of centuries. Seeing only the perfection of the completed whole, we are impressed as by some superhuman agency. But sometimes we enter such an edifice that is still partly under construction; then the sound of hammers, the reek of tobacco, the trivial jests bandied from workman to workman, enable us to realize that these great structures are but the result of giving to ordinary human effort a direction and purpose..... Science has its cathedrals, built by the efforts of a few architects and of many workers."* - Gilbert Newton Lewis, from the preface to *'Thermodynamics and the Free Energy of Chemical Substances'*, 1923. (From Coffey [8] & Hildebrand [6]).

Or his conclusion to the section on *The Future of Quantum Theory* in his monograph '**Valence and the Structure of Atoms and Molecules**', *"In that old American institution, the circus, the end of the performance finds the majority of spectators satiated with thrills and ready to return to more quiet pursuits. But there are always some who not only remain in their seats but make further payment to witness the even more blood curdling feats of supplementary performance.*

Our own show is now over and I trust that the majority of readers who have had the patience to reach this point will now leave the tent, for what I am about to say is no longer chemistry nor is it physics, nor perhaps is it sense" (From Lewis [5]).

Lewis, the experimentalist

"I have attempted to give you a glimpse...of what there may be of soul in chemistry....... Perchance the chemist is already damned and

the guardian the blackest. But if the chemist has lost his soul, he will not have lost his courage and as he descends into the inferno, sees the rows of glowing furnaces and sniffs the homey fumes of brimstone, he will call out -'Asmodeus, hand me a test-tube" (From Davenport [9]).

Lewis is remembered as one of the greatest experimentalists of all time. Like Davy whom he admired, Lewis spent most of his time in his laboratory personally conducting experiments. He did not like to spend his research conducting experiments which his chemical intuition told him would add anything new to existing knowledge. Lewis was vary of using large instruments and heavy equipment dependent research as he felt a researcher could become a slave to the instruments. He preferred to explore his inexhaustible store of ideas to develop theories which could be tested with simple experiments using easily available materials in the chemical store of the department. According to his colleagues at Berkeley he *"was ever conscious of the necessity for economy of time in research and out of the wealth of his ideas was careful to select those that would lead swiftly to the goal"* and *"the methods he chose were always simple and to the point. ..... like Sir Humphry Davy, who was one of his heroes, loved to make important discoveries with a few test tubes and simple chemicals. When the point at issue seemed to him sufficiently important, however, he would not hesitate to employ apparatus requiring skill and delicacy of manipulation, as in the beautiful but difficult experiment by which he and Calvin demonstrated the paramagnetism of the phosphorescent triplet state"* (From Gettell et al. [1]). According to Michael Kasha, Lewis loved the vacuum line set up in his work bench to *"vacuum distill liquids from one flask to another, to sublime materials into reaction vessel to watch the colour changes as the reaction took place"* (From Kasha [10]).

Lewis did not believe in an exhaustive survey of the literature in the field as he was convinced it would affect his freshness of approach and intellectual freedom. Lewis had a unique style of working. He met his students every morning and planned in detail the day's work meticulously writing the tasks to be completed by night when a review of the day's work would be done. When the research problem was completed, he would call the concerned student to the laboratory usually at night and dictate the paper to them as he paced the floor with the inevitable cigar in his mouth. The completed paper hardly needed any correction as Lewis would have thought of every detail of the paper in question (perhaps including the punctuation marks) before dictating in beautiful language.

Glenn Seaborg was a student of Lewis. He worked on Lewis acids and bases. Seaborg took pride in showing his research notebooks from those days where he had meticulously recorded his experimental work. He used to recount how Lewis would dictate research papers at night (after dinner) without much notes. He would know every detail, including the tabular matter. Seaborg later worked in other areas, especially synthesis of artificial elements (e.g., Plutonium). Seaborg used Lewis' cigar boxes to keep his rare samples. Seaborg received the Nobel Prize in chemistry in 1951.

Lewis was an undisputed mentor for a generation of great American chemists who each in his own way carried forward the Berkeley tradition of unselfishly discussing the interesting problems each was working on. In his introduction to his book on Valence, he acknowledged his indebtedness to this tradition when he wrote: *"To my colleagues and students of the University of California, without whose help this book would not have been written. In our many years of discussion of the problems of atomic and molecular structure, some of the ideas here presented have sprung from the group rather than from an individual; so that in a sense I am acting only as editor for this group"* (From Lewis [5]).

To the often asked question, could there be another Lewis? Coffey's response to the question sums up why the answer is a resounding No, "*Lewis ..... worked out the theories that explained chemistry. By the time of his death, chemical theory was firmly grounded in physics. The job of figuring the fundamental theory has already been done, so there's no opening for another Lewis*" (From DelVecchio [11]).

The end

Lewis died on Saturday the 23rd of March, 1946 in the afternoon as he no doubt would have wished – in his laboratory doing an experiment. He was found dead in the laboratory, just after he had prepared liquid HCN. Michael Kasha recalls the day vividly. Lewis was in an expansive mood in the morning as he discussed his ideas and had planned to do an experiment at his vacuum bench using liquid HCN. He went for lunch to the Faculty Club. (He seems to have met his old rival Langmuir there). Kasha who was on his way downstairs around 4 pm did not see Lewis at his bench. When he stepped in, he got a whiff of HCN gas and found Lewis sprawled on the floor, with a gash on his forehead. The vacuum line was broken and HCN liquid was bubbling in the flask. Kasha and another researcher pulled Lewis into the corridor but it was too late. Coroner's verdict was that Lewis had died of a massive heart attack. He hit his head on the bench and the impact had broken the vacuum line. Kasha firmly believed that Lewis died of a massive heart attack. There is always the question, Was it a suicide or an accident? As Kenneth Pitzer has said, Lewis was such a brilliant man, he would make a suicide look like an accident.

Tributes to an extraordinary scientist and human being

"Gilbert Newton Lewis typified the physical chemist of great intuition who was able to conceive of beautifully simple models and concepts to explain complex physical and physico-chemical phenomena.....His own career touched every aspect of science and in each he left his mark.......He epitomized the scientist of unlimited imagination and the joy of working with him was to experience life of the mind unhindered by pedestrian concerns" (From Kasha [10]).

"As a man, he Lewis was a great soul whose inspiration will never be forgotten by those who knew and loved him. He was one of those rare scientists ...., who are also great teachers and leaders of a school, so that their influence is multiplied by the many they have inspired" (From Gettell et al. [1]).

Great contributions of G. N. Lewis

- ★ Chemical thermodynamics
- ★ The Chemical bond
- ★ Acids and Bases
- ★ Heavy water
- ★ Spectroscopy and photochemistry
- ★ Chemical Education

References

1. Raymond G. Gettell, Joel H. Hildebrand, Wendell M. Latimer, G. E. Gibson, Excerpts from In memorium, 1946, University of California, Berkeley.

2. Carl S. Helrich, *Modern Thermodynamics with Statistical Mechanics*, (2009).

3. "Lewis, Gilbert Newton". Complete Dictionary of Scientific Biography. 2008. *Encyclopedia.com.* 2 Mar. 2015.

4. Norris F. Hall, *Systems of acids and bases*, *J. Chem. Educ.*, 1940, 17 (3), p 124-128, Published March 1940 & *Lewis*, J. Franklin Inst., 226, 297 (1938).

5. G. N. Lewis, *Valence and the Structure of atoms & molecules,* American Chemical Society, Monograph Series, The Chemical Catalog Co. Inc., New York, 1923.

6. Joel H. Hildebrand, *Gilbert Newton Lewis*, A biographical Memoir by National Academy of Sciences, 1958.

7. Gilbert N. Lewis, *The Anatomy of Science*, p6, Yale University Press, 1926.

8. Patrick Coffey, *Cathedrals of Science: The Personalities and Rivalries That Made Modern Chemistry,* 2008.

9. Davenport, Derek A., *Journal of Chem. Ed.*, "Gilbert Newton Lewis: 1875 - 1946", 1984, 61, p. 2. Report of the Symposium.

10. Michael Kasha, '*The Triplet State An example of G.N. Lewis research Style'*, J. of Chem Ed., volume 61, No.3, March 1984.

11. From an article ' WHAT KILLED FAMED CAL CHEMIST? / 20th century pioneer who failed to win a Nobel Prize may have succumbed to a broken heart, one admirer theorizes by Rick DelVecchio, August 5, 2006 (from the Website 'www.sfgate.com) (San Francisco Chronicle).

G. N. Lewis-Edgar Fahs Smith Collection, Schoenberg Center for Electronic Text and Image, University of Pennsylvania Library, ~1930. (From Linus Pauling -The Nature of the Chemical Bond, Special Collections & Archives Research Center, Oregon State University Libraries).

South Hall, University of California, Berkeley, 2 March 2007, Author :User:Falcorian (Wikimedia Commons).

Lewis' cubical atoms (as drawn in 1902) Author: Gilbert N. Lewis; 1902; Source: personal memorandum as found in the JFK library California State University, Los Angeles (From Wikipedia, the free encylopedia).

Gilbert Lewis, a renowned chemist at U.C. Berkeley, isolated the first sample of essentially pure heavy water from ordinary water in 1933. (Website http://www.pbs.org/wgbh/nova/hydro/water.html#h05)

16. ROBERT ROBINSON (1886–1975)

The quintessential organic chemist

"*Robinson was one of the greatest organic chemists of the 20th century. His outstanding achievements led to both a Nobel Prize and an Order of Merit, particularly for his work on the synthesis and structure of natural products, including alkaloids and plant pigments. He was also one of the founders of the electronic theory of organic chemistry*" (From Encyclopedia.com [1]).

Early years

(From "Robert Robinson", Wikipedia)

Robinson was born on September 13, 1886, to William Bradbury Robinson and Jane Davenport who was William Bradbury's second wife. William Bradbury Robinson was an innovative and enterprising man and owned a factory which specialized in manufacturing sterile surgical bandages and dressings. Young Robinson started his schooling at Chesterfield Grammar School near Leeds where its headmaster stimulated young Robert's interest in mathematics. Robert developed an abiding and lifelong interest in climbing mountains, chess and natural phenomena at work. After finishing the Grammar School, he joined Fulneck School near Leeds in West Yorkshire, a school which was known for high academic standards and teaching of mathematics. Young Robinson excelled in mathematics and physics in high school and hoped to pursue mathematics at the university but was persuaded by his father to take chemistry for his undergraduate degree at the University of Manchester as William Robinson felt that knowledge of chemistry would be more useful to the textile industry. This decision turned out to be a great boost to organic chemistry. The chemistry department at the University of Manchester was well known at that time. He obtained the B Sc degree in 1906. Later, working under William Perkin Jr, he obtained his D Sc degree from there in 1910. Robinson then became a tutor in chemistry for two years at Dalton Hall, a hall of residence at the university.

Robinson enjoyed his years at Manchester University - both from personal and professional view-points. Of the many friends he made, Robinson particularly enjoyed his friendship with Arthur Lapworth as they shared many interests such as walks in the hills, identifying wild flowers, rock climbing and playing music together (Lapworth played violin to Robinson's piano). More importantly, Robinson was impressed by Lapworth's work on chemical polarity. This laid the foundation to Robinson's work on reaction mechanisms decades later. Robinson also met his future wife Gertrude Walsh here. They were attracted to each other as they shared not only a love for chemistry but also for the mountains. They married in August 1912 and set sail to Australia soon after.

Earlier years

Robinson started his academic career in 1912 at a young age of 26 years in Australia. Encouraged by Professor Perkins, Robinson applied for and got the job as the first Professor of Pure and Applied Organic Chemistry at the University of Sydney. As he confessed later in his autobiography, Robinson took this decision as much by the attraction of New Zealand's mountains as by the career prospects. *"I applied for the post in Sydney, in the first instance, because it would be an excellent base for the exploration of the alpine chain of the New Zealand Alps. This was my primary objective, not, I am afraid at that stage, the advancement of the science of chemistry"* (From Robinson [2]). Robinsons enjoyed the openness of Australians and enjoyed a busy social life which was marred when their first child, a daughter survived only for a day. Robinson did not let this personal tragedy affect his research and worked on a number of interesting problems.

> Robinson's academic career was marked by frequent change of universities until 1930 when he finally settled down at Oxford University as the Waynflete Professor of Chemistry.

After three years in Sydney, Robinson accepted the Chair in Organic Chemistry at the University of Liverpool and returned to Britain. Though he continued his work on alkaloids, he shifted his focus to understanding how they are formed in nature. Again after just five years at the University of Liverpool, he decided to leave academia and join the British Dyestuffs Corporation as its director. The call of the academia was however, too strong to resist for long and after just a year, he moved to the University of St. Andrews, the oldest university in Scotland, as professor of chemistry (1921-1922). The short time Robinson spent at the University of St. Andrews proved to be a significant period early in his career. He published around 24 research papers from there, the paper titled *An Explanation of the Property of Induced Polarity of Atoms and an Interpretation of the Theory of Partial Valences on an Electronic Basis* published in the *Journal of the Chemical Society in* 1922 along with William Ogilvy Kermack being the most famous. Robinson used for the first time curly arrows as tools to show electron movement and charge transfer in organic reactions. His stay at St. Andrews University was also significant as he used for the first time the symbol for analogues of benzene as a hexagon with a circle in the middle in 1923. This has been universally accepted for representing analogs of benzene.

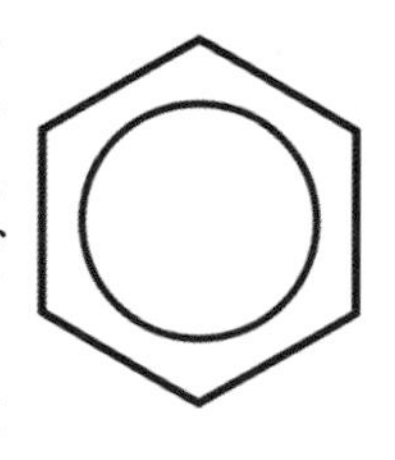

Robinson moved back to Manchester after about a year and this move was a game changer. He was able to renew his professional collaboration with Lapworth. Though he and his students continued to work on anthocyanins (plant pigments which caused red and blue colours in flowers), his most important work was on reaction mechanisms. Robinson refined the curly arrows to represent the many steps involved in a reaction mechanism. He and Lapworth used the curly arrows to pictorially represent the many steps easily to show electron movement in a reaction mechanism. (This can be compared to short hand method to record information rapidly). Robinson considered his inspirational concept of curly arrows to be at least as important a contribution to organic chemistry as his path breaking method of solving the puzzle of structure of organic compounds. This idea was accepted immediately by Ingold, an outstanding contemporary of Robinson. Ingold used the terminologies nucleophile and electrophile (instead of anionoid and kationoid used by Robinson and Lapworth) and the curly arrow to show the electron movement from nucleophile to electrophile. His terminologies and the way of using curly arrows became so universal that few chemists remembered that Robinson rather than Ingold invented the curly arrows. Robinson never forgave Ingold and felt bitter that Ingold had not acknowledged him and Lapworth as the originators of the curly arrows. In his memoirs he wrote *"I am touching on this question of acknowledgement as (Ingold) was apt to include a necessary reference to Lapworth or myself in a large number of references so that any idea that our contributions are original or applicable to the matter in hand was well and truly buried"* (From Robinson [2]).

Robinsons had another personal tragedy in 1926 when their son Michael was born with Down syndrome. Robinson moved again to accept the offer of Professorship from the University College, London, in 1928. His academic wanderings came to an end when he joined the Oxford University in 1930 as the Waynflete Professor of Chemistry where he remained till his retirement in 1955. He found the atmosphere at Oxford intellectually stimulating and challenging. However, he made very few friends during his stay there. At Oxford, though he missed the mountains, he could play challenging games of chess, his other passion, during free time.

Sir Alexander R. Todd and Sir John W. Cornforth in the Biographical memoir of Robinson have remarked *"At the chessboard, he was a deep, imaginative, and observant player,... and his power of abstract thought made him independent of visual aids to play; after his sight failed, he continued to play postal chess of a quality astonishing for a man in his eighties"* (From Halford [3]).

Why is the curly arrow so iconic?

Curly arrow can be regarded as a shorthand way of summarizing the many steps involved in a reaction mechanism and provide a quick method for drawing the changes in bonds and electrons in a given reaction. Thc ability to properly use curly arrows is perhaps the most important tool to make organic chemistry simple. They indicate how electrons move and thereby how bonds are made or broken. Curly arrows communicate non verbally with the user telling him/her clearly the bonding changes and at which particular step of the reaction the changes are taking place. The curved or curly arrow has an arrow head and a tail section. The head indicates the direction of the electron transfer and the destination of the electron and the tail shows the origin of electron. The arrow head generally has two barbs which indicate the flow of one electron pair - the electrons always moving from electron rich atom to electron deficient atom.

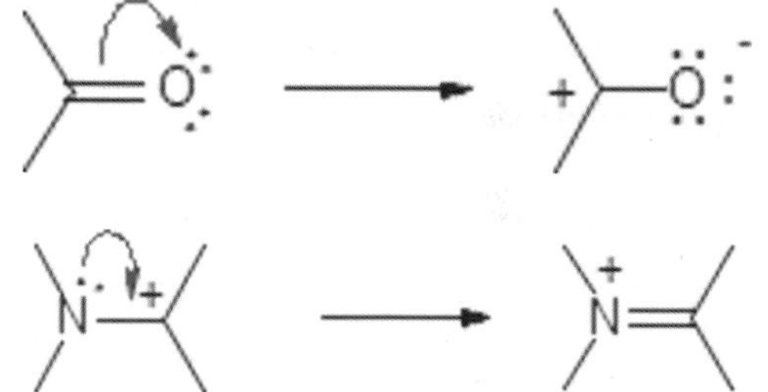

Curly arrows, organic chemist's short hand

Robinson hit upon the novel idea of using 'curly arrows' to explain elegantly and simply the movement of electrons in an organic reaction. He also found them to be a convenient tool to predict how the substitution reactions in analogs of benzene would proceed. Its simplicity and ability to pictorially convey information of electron movement in organic reactions soon gave it an iconic status and has become an integral part of organic chemist's language. They now use it routinely to describe the mechanism of a reaction or more importantly to plan the series of steps of synthesis of a new compound.

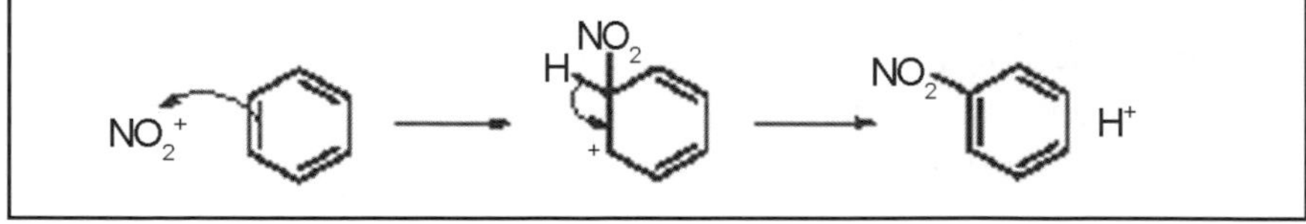

Master strategist at constructing and deconstructing molecules

Robinson was a great synthetic organic chemist. Unlike in the modern era, where organic chemists use sophisticated physical methods and computers to solve the structures of organic compounds, Robinson had to resort to first deconstructing an unknown compound into smaller and less complex pieces and then reconstruct the pieces. Robinson succeeded in determining the structure of many alkaloids using this demanding technique. This novel method of solving the structure of alkaloids (particularly of strychnine) is considered a unique example of solving the molecular structure of organic compounds. Morphine and narcotine are other molecules whose structures were unraveled by Robinson.

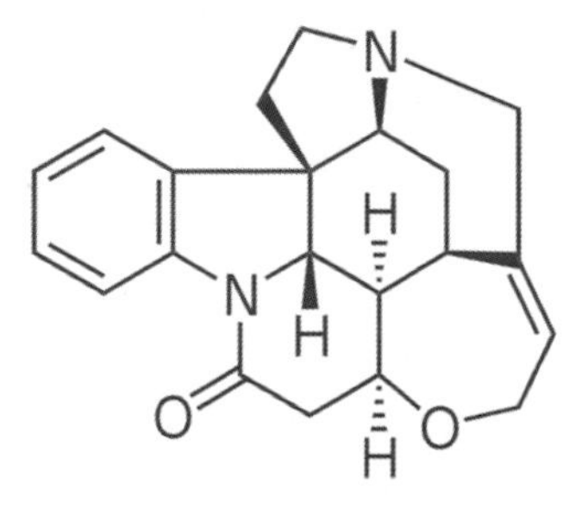

As in the case of structure, methods of synthesis of complex organic compounds was not well developed at that time. In 1917, Robinson found a unique and as yet untried method for synthesizing tropinone, an organic compound related to cocaine. He drew the complex structure of tropinone on paper, then obtained three simpler structures or building blocks of the compound by disconnecting or breaking certain specific bonds. With these figures guiding him, he succeeded in producing tropinone in the laboratory by recombining the three parts. This method of synthesis is now known as **retrograde synthesis** and is the standard procedure used in organic synthesis.

Apart from his research on the structure and synthesis of complex organic compounds, Robinson also worked on electrochemical mechanisms of organic reactions. Synthesis of alkaloids was an important area of his research. He succeeded in synthesizing many alkaloids. He was awarded the Nobel Prize in chemistry in 1947 *"for his investigations on plant products of biological importance, especially the alkaloids"* (From Nobelprize.org [4]).

Honours

Robinson received many honours and prizes from academies and universities for his path breaking contributions to organic chemistry. Some of the honours and awards he received are:, Commandeur de la Legion d'Honneur (France), Longstaff Medal (1927), Davy Medal of The Royal Society, London (1930), Royal Medal of The Royal Society, London (1932), Knight of the British Empire (1939), The Albert Gold Medal of the Royal Society of Arts (1947), Copley Medal of The Royal Society, London (1942), Franklin Medal (1947), Presidential Medal of Freedom with Silver Palm, US Government (1947), The Nobel Prize in Chemistry (1947), Priestley Medal, USA (1953), Faraday and Flintoff Medal (1960), Order of Merit (1949).

He was a Fellow of The Royal Society, London and a Foreign Member of many professional societies and academies. He was the president of the Chemical Society (1939-41) and The Royal Society of London (1945-50). He received over twenty honoris causa from various universities. He was founder of the journal, 'Tetrahedron'.

In 1962, Robinson was honoured by The Chemical Society in a unique manner. It decided to replace the conventional Presidential Address by a Robert Robinson Lectureship to be given biennially.

Robinson the man

Robinson was a man with imperious temperament and was considered to be autocratic in his dealings with other scientists. He was often dismissive of the achievements of others. His comment *"They are not chemists there, just a lot of paper hangers"* (From Birch [5]) about Frederick Sanger's use of paper chromatography is a case in point. His bitter feud with Ingold is well recorded. Robinson had a professional fight with Robert Woodward over the structure of penicillin. Woodward had used spectroscopic data to solve the structure while Robinson had no faith in this method.

Robinson was a man with varied interests. Even as an undergraduate at Manchester University, he enjoyed walking in the hills, attending music concerts, attending theatre and operas for which Manchester was quite well known. He was an avid mountain climber in his younger days. Robinson employed his gift of strategic thinking to his hobbies (particularly to solving crossword puzzles and chess). He found that solving crossword puzzles with his students sharpened his intellect. *"Each afternoon around four o'clock, just as tea was being brewed in beakers, Robinson would appear in their laboratory for a united assault on the Times crossword followed often by a discussion on any chemical topic which was currently occupying his mind"* (From Halford [3]).

His passion for chess remained throughout his life. He enjoyed playing against as many as a dozen opponents simultaneously in exhibition matches. He was the President of the British Chess Federation between 1950 and 1953. His other hobbies were photography and music.

Robinson had the idiosyncrasy of scribbling complex chemical formulae on whatever was available at that time. He scribbled on blank edges of table cloths and papers used to roll cigarette and the back of used envelopes if no writing paper was available.

Robinson's vision for organic chemistry

Robinson said in his Priestley Medal address *"We are equipped as never before, and the question arises: What shall we do with our strength?"* (From Halford [3]).

Early twentieth century witnessed rapid strides being made in spectroscopy which resulted in accelerating research in chemistry. Robinson did not have much faith in machines solving problems in organic chemistry. He chose the occasion of the Priestley Medal address to express his concern on how this could affect the future of research in the subject. Robinson visualised a future when organic chemists would employ their training both to unlock mysterious processes at work in the natural world and to mimic these processes in laboratories to create new products. Even though he himself was the master of synthesis of natural products, Robinson hoped that *"the 'marathon' or 'relay' type of synthesis will gradually fall into desuetude in the highest quarters and will be replaced by the attempt to discover how the pieces fall naturally into place. He added further that "all questions of the mechanism of reactions of carbon compounds, even in the living cell, are properly the concern of the organic chemist"* (From Halford [3]).

Robinson warned the organic chemists not to slip *"into the position of technicians, hand-maidens to other groups of scientists. My faith is that the organic chemist, as such, will initiate and develop highly significant advances in biochemistry in the next half century. I do not think that these advances can be so quickly and surely made by men trained in other disciplines"* (From Halford [3]).

Final years

Gertrude Robinson, Robinson's wife and close companion of forty two years died suddenly in 1954, just before he retired as professor at Oxford. This was a big blow to him. Robinson married again in 1957. He pursued playing chess and attending music concerts, and climbed a mountain for the final time in 1966. He lost his eyesight gradually and became practically blind towards the end of his life. He remained intellectually active and started writing his autobiography when he was 85 years old working even on the day he died on February 8, 1975.

References

1. "Robinson, Robert." Complete Dictionary of Scientific Biography. 2008. *Encyclopedia.com.* 8 Apr. 2015 <http://www.encyclopedia.com>.
2. Sir Robert Robinson, *Memoirs of a Minor Prophet: 70 Years of Organic Chemistry*, Elsevier, 1976.
3. Bethany Halford, 85th Anniversary of the Priestley Medal, *1953: Sir Robert Robinson (1886–1975)*, C&EN Special Issue, Volume 86, No.14, April 7, 2008. (quoted from (*Biogr. Mems. Fell. R. Soc.* **1976,** *22*, 414)).
4. "The Nobel Prize in Chemistry 1947". *Nobelprize.org.* Nobel Media AB 2014. Web. 9 Apr 2015. http://www.nobelprize.org/nobel_prizes/chemistry/laureates/1947/
5. Arthur John Birch, *To see the Obvious*, American Chemical Society, 1995.

17. CHRISTOPHER KELK INGOLD (1893–1970)

Founder of physical organic chemistry

"If we stand back from the detail and look down a vista of some 50 years, the conclusion is inescapable that Ingold had extraordinary imagination, insight, initiative and ingenuity, that he possessed one of the greatest intellects in chemistry and he added a new dimension to organic chemistry" (From Shoppee [1]).

"Ingold was clearly a boy genius who could turn his mind to any aspect of chemistry" (From Barton [2]).

Early years

(From www.ucl.ac.uk)

Christopher Kelk Ingold, born in Forest Gate, a suburb in the east of London, on 28th October 1893 was the son of William Kelk Ingold and Harriet Walker Newcomb. His father was a dealer in silk fabrics and suffered from tuberculosis. As both his health and his financial situation took a downturn when Christopher was just an infant, the family moved to Shanklin, in the Isle of Wight just off the south coast of England where William Ingold set up a confectionery business. The family's hope that the gentler climate would improve William Ingold's health and the new business venture were dashed as he died soon after. Ingold was five years old and his sister two years. His mother was preoccupied with the twin task of making the fledgling business to succeed and provide for the family. As a result, young

Ingold enjoyed freedom from her scrutiny and he spent much of his time exploring the flora and fauna of the island. This early exposure to nature was probably responsible for his love for outdoor activities including bird watching and rock climbing and influenced his personality as an adult.

Christopher Ingold joined secondary school at Sandown Grammar School a local county funded school on the island. Though the school was newly established, young Ingold received good education as it had enthusiastic teachers who taught well. He chose physics, chemistry and mathematics for his final two years. He qualified to appear for university admission by passing the advanced national level examination (the Intermediate BSc of the University of London). Ingold got a merit scholarship awarded by the county to study at Hartley University College in Southampton (which later became the University of Southampton). He excelled in physics there but inspired by Professor David Boyd who taught chemistry, opted for chemistry for his degree. According to Ingold, Professor Boyd was an extraordinary teacher who inspired his students by teaching chemistry as a living subject by including results of research to emphasise the exciting frontiers of the subject. Ingold confessed that he followed the same teaching strategy later to inculcate the spirit of science in his students. He obtained his B. Sc degree in chemistry from the University of London in 1913 appearing as an external candidate. He joined the Royal College of Science, a newly formed (1908) constituent of Imperial College of Science and Technology and worked with Jocelyn Field Thorpe who was a classical organic chemist. The outbreak of World War I interfered with his academic plans as he got involved in war efforts and participated in the production of an analgesic and in the development of a tear gas. He moved to Glasgow to work as a research chemist at the Cassel Cyanide Company from 1915 to 1920. Even during this period, he continued his collaboration with Thorpe and obtained the M Sc degree from the University of London in 1919.

Ingold's career can be broadly classified into three periods

Imperial college period (1920-1924) which marked the beginning of his career: The highlights of this period are his studies on tautomerism, investigation of the benzene structure, cyclisation reactions and Thorpe–

Ingold effect. He was awarded the D.Sc degree in 1921 by the University of London for the body of independent research work he had published.

Leeds period (1924-1930) was when Ingold changed the traditional approach to organic chemistry: This can be regarded as the incubation period during which Ingold metamorphosed into his final avtaar of a physical organic chemist (Ingold probably would have objected to this nomenclature later) using tools of physical chemistry to solve problems of reaction mechanisms of organic compounds. Here he studied substitution, addition and elimination reactions, theories of electronic effects, ester hydrolysis, tautomerism and mesomerism.

University College London period (1930-1961): This was the most prolific period of Ingold's professional life. This period also witnessed Ingold's legendary research collaboration with Hughes. In addition to his outstanding research in a number of areas, his contribution as the head of the chemistry department is monumental.

During 1921-24, Ingold published 42 papers (mostly in classical organic chemistry and a few in physical chemistry). He was the first recipient of the newly established Meldola Medal given to a chemist under thirty years of age in recognition of the outstanding research done during the year. 1923 was special on two counts - he got the Meldola medal for the second time and was elected a fellow of the Royal Institute of Chemistry. His research output was so prolific that by the end of 1924 in a short span of four years, he had published 54 papers. With Thorpe, his mentor and guide at the University of London, he also published a book on synthetic dyes. In the same year he was elected to the Royal Society when he was just 30 years old (one of youngest to be elected a fellow of the prestigious Society) in recognition *of "not only his productive experimental work in organic chemistry and an innovative and promising mathematical approach to the subject, but also important papers published in the field of physical chemistry"* (From Encyclopedia.com [3]). The year 1924 also saw his appointment to the chair of organic chemistry at Leeds University where he remained for the next six years and assembled a large research group. Thus began Ingold's extraordinary research career.

This move to Leeds helped Ingold as it was here that he began his career as a physical organic chemist. Ingold came under the influence of Professor Harry Medforth Dawson (professor physical chemistry) and began to get more interested in physical chemistry and particularly in kinetics of organic reactions. It marked the beginning of Ingold's research interest in the study of the reaction mechanisms and kinetics of substitution and elimination reactions. This novel method of understanding the subject of organic chemistry became the accepted route in both research and teaching of organic chemistry.

Ingold moved to University College London in 1930. One of the first tasks he had to undertake was building his research group afresh as only one student had come with him from Leeds. Ingold was looking for someone with a good grasp of kinetics of reaction mechanisms of organic compounds. He remembered that as the PhD examiner of Hughes, he had been impressed by Hughes' understanding of chemical kinetics. The unique research collaboration between Ingold and Hughes started when Ingold persuaded Hughes to join him at the University College London as his post-doctoral assistant. This was the beginning of one of the most remarkable and long lasting research collaborations which went on till Hughes' death due to cancer in 1963. Hughes' mastery over chemical kinetics proved to an ideal foil to Ingold's expertise in electronic theory and they complemented each other with their individual expertise blending perfectly. In their long research collaboration, they published around 140 papers (Ingold published a total of around 280 papers during that period) and Hughes published a total of 230 papers. Their major collaborative work was the classification of the reaction mechanisms of aliphatic compounds. They used the novel approach of employing tools and methods of physical chemistry to understand how variations in structures and the conditions under which the reactions occurred influenced and changed the way in which the reactions progressed. They published some of their research findings in Nature (From Hughes & Ingold [4]). They also found solutions to many vexing puzzles of classical organic chemistry. Barton regards their *"revelationary clarification of the hitherto mysterious Walden Inversion"* (From Barton [2]) which had baffled classical organic chemists for four decades as one of their *"noteworthy accomplishments"*

(From Barton [2]). Until now *"organic chemists rarely made kinetic studies on preparatively significant organic reactions at that time. Hughes and Ingold opened a new era of precision"* (From Barton [2]).

Their idea that the basis of all branches of chemistry including organic chemistry was physical chemistry did not find acceptance by classical organic chemists. It took all of Ingold's persuasive and communication skills apart from evidence of research findings to overcome this resistance. Ingold took a sabbatical leave at Stanford during 1932 and while there he wrote a 50 page comprehensive summary on the principles of the electronic theory of organic reactions and published it in *Chemical Reviews* of the ACS. He expressed his conviction that *"the new work made it inescapably clear the old order in organic chemistry was changing, the art of the subject diminishing, its science increasing: No longer could one just mix things; sophistication in physical chemistry was the base from which all chemists, including the organic chemist, must start"* (From Barton [2]). While the article, written in simple language consolidated Ingold's reputation in the field, it precipitated the clash between Ingold and Robinson, the two titans of the electronic theory of organic reactions. He also explained the term 'mesomerism' which was similar to Linus Pauling's resonance concept in the article. He had used it in 1924 itself but had not explained the concept. Ingold acknowledged Hughes's invaluable contribution in bringing about the fundamental change in organic chemistry's approach to problems declaring *"this work has changed the aspect of organic chemistry, by progressively replacing empiricism by rationality and understanding, to a degree which is now manifest in the terminology and teaching of the subject, and in the research activity all along its advancing frontier"* (From Ingold [5]). This statement applies equally to Ingold as well.

Structure of Benzene

During the golden decade of the 1930s, Ingold revisited his earlier work on solving the structure of benzene molecule. He realized that the best approach was to use a number of new techniques that were available at that time to solve this decades old problem. Hughes, his collaborator had just then set up a plant to produce deuterium atoms and this gave Ingold the idea of substituting hydrogen atoms with deuterium atoms in the benzene molecule and successfully synthesized deuterated benzene C_6D_6. He applied the quantum mechanical calculations of Edward Teller (visiting professor at University College London, 1934-1935) and concluded that the benzene molecule had a planar hexagonal structure which had been predicted by Wilson & Pauling. Ingold published eight papers about this work in the *Journal of the Chemical Society* in 1936 and completed with a series of papers in 1946. An entire monthly issue of the Journal of the chemical society was dedicated to Ingold's work on this subject.

Stereochemistry

During 1950s and 1960s Ingold collaborated with Robert S. Cahn and Vladimir Prelog and developed a set of sequence rules to assign specific positions for the four different groups bonding with the tetrahedral chiral carbon atom. This sequence rule came to be known as Cahn-Ingold-Prelog or simply CIP rule and in the course of time became an internationally accepted rule used to find out if an optically active molecule was R (rectus/ right handed) or S (sinister/ left handed) (From Cahn, Ingold & Prelog [6]).

Ingold delivered the Bakerian lecture of the Royal Society in 1938 on *The Structure of Benzene* and the Baker lectures at Cornell university during 1950 and 1951. Based on these lectures, he published in 1953 what many regard as his magnum opus, his famous eight hundred page classic book *Structure and Mechanism in Organic Chemistry* intended for university students. He published over 450 research papers in prestigious scientific journals.

A new approach to organic chemistry

"Christopher Kelk Ingold, the 'boy genius' who grew up to be an extraordinary chemist forced a fundamental change in organic chemistry in an era when the world of science was changing" (From Leffek [7]).

Organic chemistry prior to Ingold's intervention had essentially concentrated on classification of carbon compounds on the basis of their structures and properties. Ingold was convinced that a better way to understand organic chemistry was through applying principles of physical chemistry and using its tools to understand reaction mechanisms and electronic structure of carbon compounds. He was able to give a new direction to the study of organic chemistry and systemize the field. He established some of the fundamental reaction mechanisms of organic chemistry, especially the mechanism of substitution reactions. He was the first to describe two different forms of substitution reactions, S_N1 and S_N2. He was an excellent communicator of his ideas and convinced the nonbelievers by introducing new terms to explain his novel ideas. He introduced terms such as nucleophile, electrophile, S_N1, S_N2, E1, and E2 to describe the nature of the chemical reactions into the vocabulary of everyday usage of organic chemistry. He also introduced the term mesomerism in 1924 to describe resonance effects.

Department of chemistry at the University College London (UCL)

"Ingold had little regard for division of chemistry into subdisciplines, because he saw the subject as a whole and as part of the grand pattern of science" (From Bunton [8]).

"A revolution has occurred in chemistry in my life-time; and it is continuing and cannot be resisted. When I began, chemistry was almost wholly a mass of empirical observation with a little regularity, but without either reason or coherence. Today however, the outlook is quiet changed. The whole of chemistry is bound together and rationalized by physical principles..." (From Roberts [9]).

"It is part of this development that barriers between conventional divisions of chemistry are breaking down to the point at which ......there are hardly any problems assignable to one division which cannot be assisted by the ideas associated with another" (From Roberts [9]).

When Ingold joined UCL, Donnan was the head of the department. He was committed to the view that artificial subdivisions of chemistry as physical, organic and inorganic hindered the progress of not only chemistry but also of science itself. Furthermore, he was convinced that interpreting chemistry through theories and methods of physical chemistry would contribute to the development of the subject in future. Ingold also believed in the same concept of chemical science and worked towards this after he succeeded Donnan as the head in 1937. One of the ways to achieve this objective was the unwritten rule that all students and researchers of the chemistry department had to attend lectures and colloquia of not only senior faculty members and visiting scientists but also of students presenting various aspects of their research. He firmly believed that this would result in 'cross fertilization' of ideas and would enormously benefit the scope of chemistry. Ingold succeeded in reforming the B.Sc course changing it from a two year course to a three course and making two years of physics and one year of mathematics course work mandatory. By 1948, University College London was like a magnet to chemists from all parts of the world and had become the centre championing the unity of chemistry.

As a teacher: Ingold was a brilliant teacher. *"His teaching aids were chalk, chalkboard and an incisive mind……His classroom lectures were elegant intellectual presentations which forced his audience to think, and keep on thinking, and after the class he displayed infinite patience in dealing with questions"* (From Bunton [8]). His patience in dealing with young students was well known. Once a young female student arrived a little late for his lecture. Ingold instead of getting annoyed stopped speaking, patiently waited for her to get settled and then said, *"what I have just dealt with is rather important and so I should perhaps start again"* (From Ridd [10]).

Ingold taught chemistry as a living science constantly expanding and evolving with exciting prospects. He included anecdotes of recent research in the subject and inculcated the spirit of science. He believed that in the fast changing world, a minimum scientific knowledge was an essential requirement for all. He was convinced that those who were engaged in doing good science should take the extra responsibility to communicate the wonder and excitement of doing science through lectures. *"His lectures were indeed like sermons. He spoke with wonderful clarity and precision. He started with the proposition, followed by evidence.........but all was illuminated as we approached the beautiful revelation of truth"* (From Barton [2]).

Honours

Ingold received the Davy Medal of the Royal Society (1946), Longstaff Medal (1951) of the Royal Society of Chemistry, The Royal Medal of the Royal Society in 1952 and knighthood in 1958. He received the first Flack Norris Award of the ACS (1965). He also received honorary doctorates from numerous universities. Ingold was the president of the Chemical Society (1952-54). He was honoured by UCL by establishing Sir Christopher Ingold building in 1969 locating its chemistry department there. The Royal Society of Chemistry established the Christopher Ingold Lectureship in 1973 to commemorate his enormous contributions. The Ingold lectureship was discontinued in 2008 as a stand alone award but continues as part of the physical organic chemistry prize of the Royal Society of Chemistry.

One prize eluded Ingold – The Nobel Prize. Many have speculated that perhaps this was the price he paid for his clash with Sir Robert Robinson over the electronic structure and the famous or infamous 'curly arrows'. Barton [2] and many others subscribe to this view. Ingold joined the club of distinguished chemists who did not get the PRIZE.

Ingold the man

Ingold was a generous and a genial person who got along well with fellow scientists. Many who knew him felt that *"it was very difficult to separate Ingold the scientist from Ingold the individual because of his overwhelming devotion to science"* (From Bunton [8]). His commitment to people was as total as his commitment to science. This is best illustrated by his acute awareness of his responsibility to young scientists trying to establish themselves. He did not put his name on papers published by young colleagues even when the ideas had originated from him. He set up a fund to help needy scientists to overcome their financial difficulties and the trauma of the world war. The beneficiaries often were unaware of who their godfather was. Ingold was proactive in helping those fleeing from Germany during the World War period. His was an open house for anyone needing shelter. His son Keith has recalled how he had to give up his room when people arrived unexpectedly (From Ridd [10]). Ingold helped innumerable foreign students financially from his private funds until they could support themselves.

At a personal level, he was happily married to Edith Hilda Usherwood, Hilda to friends. They met when she came to study chemistry at Imperial College. She worked with Ingold for a few years but gave up her profession and devoted herself in assisting her husband. Interestingly, Ingold's hobbies were uncannily similar to those of his bitter rival Robinson. He loved mountaineering and bird watching. While at Leeds he regularly climbed the steep rock faces of Yorkshire hills. One of his great regrets was that he had not climbed the Matterhorn. He kept a model of this mountain on his desk as if to remind himself of one of his unfulfilled tasks. He enjoyed playing chess but not with Robinson's intensity. He apparently confessed to C W Shoppee, his biographer that he would have got better grades if he had spent less time in the student union at the Hartley College playing chess.

Ingold was a colossus in his chosen field and strode across the different aspects of chemistry and left a far reaching legacy of an unified approach to the subject and permanently changed the approach to organic chemistry.

He passed away on 8 December 1970 after a brief illness.

References

1. Charles W. Shoppee, '*Christopher Kelk Ingold.1893-1970*', Biogr. Mems Fell. R. Soc. 1972, 18, 348-411.
2. Derek H. R. Barton, *"Ingold, Robinson, Winstein, Woodward, and I"* Bull.Hist.Chem. 19 (1996), pp (43-47).

3. "Ingold, Christopher Kelk." Complete Dictionary of Scientific Biography. 2008. Encyclopedia.com. (April 24, 2015). http://www.encyclopedia.com.

4. E. D. Hughes and C. K. *Ingold "Dynamics and Mechanism of Aliphatic Substitutions"* Nature 1933, 132 (933-934).

5. Christopher K. Ingold, *'Edward David Hughes 1906-1963', Biographical Memoirs of Fellows of the Royal Society,* 10 (1964): 147–182).

6. R. S. Cahn, C. K. Ingold and V. Prelog, *Specification of Molecular Chirality*, Angewandte Chemie, 1966, 78, 8, 413-447.

7. Leffek, Kenneth T., *Sir Christopher Ingold: A Major Prophet of Organic Chemistry*. Victoria, BC, Canada: Nova Lion, 1996.

8. Clifford A. Bunton, *C. K. Ingold: A chemical revolutionary*, Pure & Appl. Chem., Vol. 67, No. 5, pp. 667-672, 1995.

9. Roberts K. Gerrylynn, *C. K. Ingold at University College London: Educator and Department Head,* Bull. Hist. Chem. 19 (1996).

10. John H. Ridd, *Organic Pioneer*, Chemistry World, 2008, pp. 50-53. (www.chemistryworld.org).

1937 Sir Christopher Ingold, Retrieved from http://www.ucl.ac.uk/chemistry/history/chemical_history/slides/1937.

18. HENRY EYRING (1901–1981)

A simple man with an active mind

"Great ideas come from simple people. It is simple ideas that can actually change the world. Henry Eyring instinctively knew the truth when he saw it. You know .... it. The truth is always simple. The lesson of Henry Eyring's life is that simple people, just like you and me, can change the world. We do it a little bit every day. And we have the potential to change the world much more, if we can better understand and use our unique gifts" (From Dambrowitz & Kuznicki [1]).

"Henry Eyring was fortunate in entering the arena of chemical physics at the time that quantum mechanics began impinging on the fundamental problems of chemistry. He was also fortunate in possessing to an unusual degree a fertile imagination, unbounded curiosity, a warm and outgoing personality, a high degree of intellectual talent, the ability to work hard, and a determination to succeed. The result was that, beginning in the early years of the 1930s, he exerted an important influence on the large numbers of students and colleagues lucky enough to come into contact with him. This influence continued to spread throughout the chemical community for the rest of his life. He broke new ground in a wide sweep of scientific activities, involving matters that ranged from fundamental principles of chemistry to problems of a highly practical and applied nature" (From Kauzmann [2]).

(From famousmormons.net)

"For me there has been no serious difficulty in reconciling the principles of true science with the principles of true religion, for both are concerned with the eternal verities of the Universe. . . ." (From Eyring [3]).

Early years

Henry Eyring, the son of Edward Christian Eyring, a second generation practicing Mormon, and Caroline Romney Eyring was born on February 20, 1901 in a Mormon settlement, Colonia Juárez, in Chihuahua, Mexico. Edward Christian Eyring was a member of The Church of Jesus Christ of Latter-day Saints or LDS Church for short. His father had two wives which his religion allowed, and until 1954 lived in a polygamous marriage when one of his two wives died. Henry Eyring was the eldest of the 18 children. The children were brought up according to Mormon teachings. Importance of hard work and equality of all human beings were impressed upon them from childhood. This strict religious upbringing influenced Henry's attitudes in later life as a scientist. When he became a famous scientist, Henry was often asked if there was any conflict between science and religion? His simple reply was *"There is no conflict in the mind of God, but often there is conflict in the minds of men"* (From Eyring [3]).

Henry Eyring and siblings had a carefree childhood on the large cattle ranch in Chihuahua, Mexico. When he was eleven years old Eyrings had to abandon all their assets and flee from Mexico along with some 4000 odd immigrants when the Mexican revolution broke out in July 1912 and the prosperous American immigrants became targets of the insurgents. Eyrings first settled down in Texas (El Paso) across the border but moved to Arizona and settled in Pima in 1914. Henry Eyring joined high school there and later Gila Academy in Thatcher, Arizona (now called Eastern Arizona College). At school, he enjoyed mathematics and science as he discovered he had an aptitude for these subjects. Henry received a fellowship to do graduate study at the University of Arizona, from where he obtained Bachelor's degree in mining engineering in 1923 and Master's degree in Metallurgy in 1924. His practical experiences in the underground mines and the smelting of ores during his master's program, made Eyring choose chemistry for his PhD degree.

> The story behind the choice of the subject of study reveals the practical side of Henry's character. It is said that while he was working in a mine, a large rock fell on his boot causing a painful injury. This made Eyring think of the frequent injuries – often fatal- suffered by miners, and decided against taking up mining as a career. He ruled out metallurgical engineering as a career possibility because of the toxic fumes from the smelter. This left chemistry as the choice (From Eyring [4]).

Eyring joined the University of California, Berkeley, for his PhD. Though he was not formally associated with G. N. Lewis (his formal guide was Professor George Ernest Gibson), Eyring like all others in the chemistry department at Berkeley was strongly influenced by G. N. Lewis. He shared Lewis' view of physical chemistry (*"Physical chemistry is everything that is interesting"* (From ACS [5])). Eyring employed his considerable skills and expertise in physical chemistry to solve problems from diverse fields including biology. Eyring was awarded the PhD degree in 1927 for his thesis *"A Comparison of the Ionization by, and Stopping Power for, Alpha Particles of Elements and Compounds"* (From "Henry Eyring (chemist)," *Wikipedia* [6]).

Early career

After getting his PhD degree, Eyring joined the University of Wisconsin in 1927 as a postdoctoral research fellow for two years. In 1929, Eyring got a national research fellowship for a year and decided to work with Polanyi at Kaiser Wilhelm Institute in Berlin. During this period Eyring and Polanyi conducted some of the earliest studies on reaction dynamics for elementary bimolecular gas-phase chemical reactions. They developed a method by which empirical results substantiated by theoretical calculations could be used to calculate approximately the rate of a chemical reaction. Eyring returned to the US in 1930 and after a year at Berkeley, joined Princeton University in 1931. This marked the 15 year long and productive association with the Chemistry Department at Princeton University.

Absolute Rate Theory

Within a short period of four years, Eyring published his radical idea of Absolute Rate Theory (ART) in 1935 in the journal of Chemical Physics (From Eyring [7]) and followed this up a year later with another classic paper on the subject (From Eyring [8]). The ART idea is one of the most influential developments of the twentieth century and has had a tremendous impact on chemistry. Eyring was proud of this work and explained its importance thus *"I showed that rates could be calculated using quantum mechanics for the potential surface, the theory of small vibrations to calculate the normal modes, and statistical mechanics to calculate the concentration and rate of crossing the potential energy barrier. This procedure provided the detailed picture of the way reactions proceed that still dominates the field"* (From Eyring [9]). According to this theory the rate of a chemical reaction is controlled by the definite mean lifetime of the activated complex which is definite and does not vary. In simpler words, according to ART, *"atoms and molecules can collide and combine to form an unstable, high energy complex. When the molecules fall out of this high energy state, they may do so as new and different molecules, or in their original states. The energy required to reach the activated state must be available if the molecules are to change into something new"* (From Dambrowitz & Kuznicki [1]).

Eyring's transition state theory has been used by innumerable scientists and was the forerunner of later developments in chemical dynamics.

The paper on ART was rejected twice but Eyring was convinced that he was correct in his surmise and did not abandon the idea. When his paper was rejected the first time, Eyring had a near fatal accident. On coming out of it, he took it as an omen that he should not give up the idea! His perseverance paid off when the theory was finally accepted and became one of the most important theories in chemistry. He was able to make this quantum jump as he was prepared to go beyond the accepted paradigms and was willing to use his imagination and think out of the box.

This seminal work catapulted Eyring into scientific limelight but he remained as before the humble and hard working scientist, enjoying teaching the introductory chemistry course through which he passed on his love for chemistry to his young students.

While he was at Princeton University, Eyring co-authored with J. Walter, and G. E. Kimball, the book *Quantum Chemistry* (published in 1944). This book soon came to be regarded as the standard text on the subject. In the introduction to the book he wrote *"No chemist can afford to be uninformed of a theory which systematizes all of chemistry even though mathematical complexity often puts exact numerical results beyond his immediate reach"* (From Eyring, Walter & Kimball [10]). Eyring moved to University of Utah as professor of chemistry when he accepted the offer of deanship from the University. In many ways this move was like home coming to Eyring as Utah had a sizable Mormon population and a very active LDS Church. Eyring who was already a famous scientist became the science face of the church.

Eyring's approach to science

"I am a dedicated scientist and the significant thing about a scientist is this: he simply expects the truth to prevail because it IS the truth. He doesn't work very much on the reactions of the heart. In science, the thing IS, and its being so is something one cannot resent. If a thing is wrong, nothing can save it, and if it is right, it cannot help succeeding" (From Eyring [3]).

He was driven by childlike curiosity and was *"more interested in discovering what is over the next rise than in assiduously cultivating the beautiful garden close at hand"* (From Eyring [9]). Eyring did not waste his energy in worrying about minute details of a problem and instead went to the heart of the problem. He believed in the advice he gave to his son Hal that *"You ought to find something that you love so much that when you don't have to think about anything, that's what you think about"* (From Eyring [4]). He worked as hard or harder than anyone, which is part of the reason he finished his Ph.D. thesis work in two years. He believed that *"A scientist's accomplishments are equal to the integral of his ability integrated over the hours of his effort"* (From Hirschfelder [11]). Even after he became famous, Eyring was not afflicted by self-importance and did not take anything for granted. He believed in working harder than others because he often said *"if the economy goes to ruin and there's only one chemist in the country with a job, it's going to be me"* (From Eyring [4]). Throughout his long academic career, he was always proud that he "taught a full load" of teaching six days a week. A year before he died, at the official function of renaming the chemistry building as Henry Eyring building, he responded to the honour with a typical speech showcasing his work ethics, humility and humour *"I'll keep working as long as I can find my way to the chemistry building and somebody there will let me in. Now that my name is on the building, it should be a lot easier"* (From Eyring [4]). True to his word, during 1980 -81, Eyring taught a full undergraduate course, pursued active research and was an active collaborator of three books (From Henderson [12]).

Henry Eyring believed that something he wanted badly could be achieved if he strove hard enough. A typical example of this is that while at high school inspite of his height (or the lack of it, for he stood at 5 feet 8 inches), he played as center of his high school basketball team in every game (without being replaced during the game). The team won the area championship.

Another facet of his personality was that undaunted by the innuendoes by some of his colleagues, he did not abandon his unique approach to scientific problems which was driven by curiosity. On the other hand, he *"gave folks courage to be very innovative, to think broadly about scientific problems, to speculate and be willing to cross disciplinary boundaries......He went where his curiosity took him"* (From Eyring [4]).

Eyring's insatiable curiosity about things around him made him interested in finding the answer to whatever caught his attention. This trait made him different from other great scientists. The story that Eyring himself was fond of recounting best illustrates this difference. Once Eyring and Einstein were taking a walk in Princeton through what was a rose garden but was replanted with field crops. Eyring was intrigued by the new crop and asked Einstein if he had any idea. Einstein did not, and continued to concentrate on the problem they were discussing. To satisfy his curiosity, Eyring approached the gardener and found out it was soya beans.

His curiosity driven approach enabled Eyring to be open to new ideas and he was not afraid to think innovatively to explore new problems that appealed to his imagination. The fallout of this was, apart from his outstanding work in physical chemistry, Eyring collaborated with scientists across disciplines and published papers in such diverse fields as biochemistry, biology, astrophysics, geology, medicine, molecular biology. He was particularly interested in chemical education. He believed that even the most complex concept or idea can be taught effectively if the teacher has a deep and clear understanding of its essential ideas. His principle was *"If you can't explain something to an eight-year-old, you don't really understand it yourself"* (From Eyring [4]).

Honours

Henry Eyring was one of the most decorated chemists. Among the honours and medals he received are: Newcomb Cleveland Prize of the American Association for the Advancement of Science (AAAS) (1932), Bingham Medal of the Society of Rheology (1949), Peter Debye Award in Physical Chemistry (1964), the National Medal of Science (1966), Irving Langmuir Award (1967), Linus Pauling Award (1969), Elliott Cresson Medal of the Franklin Institute (1969), T. W. Richards Medal (1975), Priestley Medal (1975), Berzelius Medal of Sweden (1979), (Eyring was particularly proud of the Berzelius medal as it is awarded once in 50 years), Wolf Foundation Prize of Israel (1980).

He was elected to the National Academy of Sciences (USA) in 1945.

He was the president of the American Chemical Society in 1963 and of the American Association for the Advancement of Science in 1965. However, the Nobel prize eluded him.

Eyring, the man

Eyring was childlike. He was by nature, trusting, without guile and transparent in his dealings with people and above everything else, curious. *"I perceive myself as rather uninhibited, with a certain mathematical facility and more interest in the broad aspects of a problem than the delicate nuances"* (From Eyring [9]). He had a prolific publication record. When someone asked him why he wrote so many papers, he is said to have remarked *"I suppose it's some kind of egotism. I like to understand what molecules do, but I also like to tell others what I know"* (From Eyring [4]).

Eyring believed in the biblical saying do unto others as you wish them to do unto you. He emphasized the *"importance of being good to people you pass on the way up because you will want them to be good to you when they pass you on their way up and you are on the way down"* (From Eyring [4]). He was always courteous and never took a confrontational approach even to those who criticized him and held contrary views. That did not mean that he would meekly give up his convictions. According to his grandson Henry J. Eyring, he would say, *"Let's talk together about where the truth really lies. Let's learn together"* (From Eyring [4]).

Eyring believed that in the eyes of God all are created equal and therefore everyone should be treated with courtesy and respect. There is an inspiring story of this facet of Eyring's personality. When he was the Dean of the graduate School at the University of Utah, Brother Amott, the ever courteous mailman delivered mail to the Dean's office twice daily without fail for twenty years. Eyring decided to honour the mailman by awarding Brother Amott with an honoris causa degree in recognition for his selfless work (From Dambrowitz & Kuznicki [1]).

Eyring and his religious belief

Eyring was a deeply religious man. He found no conflict in being a scientist and religious person at the same time. *"Contemplating this awe-inspiring order extending from the almost infinitely small to the infinitely large, one is overwhelmed with its grandeur and with the limitless wisdom which conceived, created and governs it all. Our understanding, great as it sometimes seems, can be nothing but the wide-eyed wonder of the child when measured against omniscience"* (From Eyring [3]).

According to Eyring, the apparent contradictions between religion and science, *"are to be expected as long as human understanding remains provisional and fragmentary. Only as one's understanding approaches the divine will all seeming contradictions disappear"* (From Eyring [3]).

In October 1981, a scientific meeting was organized in Berlin to commemorate the fiftieth anniversary of the famous paper "Über einfache Gasreaktionen." by Henry Eyring and Michael Polanyi. Two months later Eyring died on 26 December 1981 in Salt Lake City at the age of 80, working till the last day of his life.

The best tribute one can pay this extraordinary scientist is to remember his exhortation *"May we live and understand it in a big way and not worry about the small things that we do not understand very well, because they will become clearer as we go on"* (From Eyring [3]).

References

1. K. A. Dambrowitz and S. M. Kuznicki, University of Alberta, *Henry Eyring: A Model Life*, Bull. Hist. Chem., Volume 35, Number 1, November 2010.
2. Walter Kauzmann, *Henry Eyring (1901- 1981);* A Biographical Memoir, National Academy of Sciences, National Academies Press, 1996.
3. Henry Eyring, *The faith of a Scientist*, Bookcraft, 1967.
4. Henry J. Eyring, *Mormon Scientist: The Life and Faith of Henry Eyring,* Deseret Book, 2007.
5. "Physical Chemists, Explore The Way Things Work." American Chemical Society, 2008.
6. "Henry Eyring (chemist)." *Wikipedia, The Free Encyclopedia.* 5 May. 2015. Web. 14 May. 2015.

7. H. Eyring, *"The Activated Complex in Chemical Reactions"* J. Chem. Phys., 1935, *3*, 107-115.

8. H. Eyring, "*Viscosity, Plasticity and Diffusion as Examples of Absolute Reaction Rates*," *J. Chem. Phys.*, 1936, *4*, 283-291.

9. H. Eyring, *Men, Mines, and Molecules*, Ann. Rev. Phys. Chem. 1977, 28, 1-13.

10. H. Eyring, J. Walter, and G. E. Kimball, *Quantum Chemistry*, Wiley, New York, 1944.

11. J. O. Hirschfelder, "*A forecast for Theoretical Chemistry*", J. Chem. Educ., 1966, 43 (9), 457-463.

12. D. Henderson, "*My friend, Henry Eyring*", J. Phys. Chem., 1983, 87, 2638-2656.

Transition state and the Nobel Prize

Eyring's transition state theory has been invoked and used by a large number of chemists to understand kinetics and mechanisms of reactions. The early transition state theory has undergone much change over the years.

Eyring did not get the Nobel Prize, but a number of chemists have received the Nobel prize for studies in chemical dynamics.

1956	C. N. Hinshelwood and N. N. Semenov (Mechanisms of reactions)
1967	M. Eigen, R. G. W. Norrish and G. Porter (Fast chemical reactions)
1986	D. R. Herschbach, Y. T. Lee and J. C. Polanyi (Dynamics of chemical elementary processes)
1992	R. A. Marcus (Theory of electron transfer)
1999	A. H. Zewail (Transition states of chemical reactions using Femtosecond spectroscopy)

19. LINUS PAULING (1901–1994)

The irrepressible scientist and crusader, with two unshared Nobel prizes

"A multifaceted genius with a zest for communication" (From Marinacci [1]).

There was one Pauling and there will be no other.

"Linus Pauling for years was probably the most visible, vocal, and accessible American scientist. A black beret worn over a shock of curly white hair became his trademark, along with a pair of lively blue eyes that conveyed his intense interest in challenging topics. He was a master at explaining difficult, even abstruse, medical and scientific information in terms understandable to intelligent lay persons" (From Marinacci [1]).

Early years

(From Nobelprize.org)

Linus Pauling was born 28 February 1901 in Portland, Oregon, to a self-taught druggist, Herman Henry William Pauling and Lucy Isabelle (Belle) Darling, the descendent of a pioneer family. Pauling's parents came from contrasting backgrounds. Herman Pauling's ancestors were immigrant Germans known for their sobriety and strong work ethics. The Darlings were eccentric and enterprising.

Linus *"had a grandfather who practiced law without a degree; a great uncle who communed with an Indian spirit; an aunt who toured the state as a safecracker (legally; she practiced her skills for a safe company); and a mother whose chronic anemia kept her bedridden for long stretches"* (From Mead & Hager [2]).

The Paulings lived in poverty as Herman did not have a steady job. Herman constantly moved from place to place and job to job resulting in an unsettling atmosphere at home. The family finally settled down at Condon, an arid small town in interior Oregon. Herman took great interest in Linus and encouraged him to develop the reading habit. When interviewed about his relationship with his father, he remembered his father as *"a kind and caring man who protected his family, even at cost to himself"* (From Hager [3]). Linus grew up in Condon where hard work was appreciated and *"people were valued for the work they did, not the name they carried"* (From Mead & Hager [2]). Linus spent time at his father's drugstore where he befriended both cowboys and Indians. He learnt *"the proper way to sharpen a pencil with a knife"* from a cowboy, and *"how to dig for edible roots from an Indian"*. Linus recalled later that these two experiences made an indelible impact on him as they showed him *"there was correct technique for doing things and that there were people who had useful knowledge of nature"* (From Paradowski [4]). Tragedy struck Linus at age nine when his father who had been suffering from ulcers died leaving his family with no financial support. Belle Pauling, stunned by her husband's sudden death and suffering from severe anemia, was obliged to run a boarding house on the outskirts of Portland to support her children. Nine year old Linus, who had become a voracious reader by then, felt the loss of his father keenly and found solace in books and hobbies. He had to take up odd jobs while still young to augment the family income.

When Linus was 14 years old, a visit to his friend Lloyd Jeffress who had set up a small chemistry laboratory in his bedroom changed his life completely. Linus was fascinated by chemical reactions which brought about strange changes in the substances and the smoke and smell that accompanied these changes. (Jeffress did the reaction between sulfuric acid and a mixture of potassium chlorate and sugar). Pauling scraped around and set up a chemistry laboratory in the basement of his house and found peace experimenting in his home laboratory.

Linus Pauling joined Washington High School in Portland in 1916. The chemistry teacher there encouraged his interest in the subject. He excelled in science and mathematics but did poorly in American History. As a high school senior Pauling had enough credits to get admission to the Oregon Agricultural College at Corvallis (now Oregon State University) but did not have credits in American history courses to get his formal high school diploma. Linus did not wait to get his high school diploma as the school authorities refused permission for Pauling to complete course requirements concurrently with his university work. (It took 45 years and two Nobel Prizes for the school to relent and award him the high school diploma). Linus used his savings to finance his studies and joined the chemical engineering department as a freshman in 1917 Fall term. Pauling was so confident of his ability to handle the heavy academic load that he took 9 courses – two courses each in chemistry and mathematics in addition to 5 other courses.

"Only when I began studying chemical engineering at Oregon Agricultural College did I realize that I myself might discover something new about the nature of the world" - Pauling (From Marinacci [5]).

Linus was forced to take a break from his junior year in 1919 to supplement his mother's meager income. He took up various jobs, often paving roads for long hours, both to support his family and save up enough to rejoin the university. Pauling got a respite from financial worries when he was offered teaching assistant's job to teach quantitative analysis at his college. Pauling's office as a teaching assistant was located in the chemistry library and he began reading scientific journals during his spare time. Pauling came across G. N. Lewis' seminal paper of 1916 on how chemical bonds are formed by the sharing of electron pairs and Irving Langmuir's papers on the electronic theory of valence. These papers triggered Pauling's life-long interest in chemical bonds. Pauling himself remarked later *"I had become interested in the question of the nature of the chemical bond, after having read the 1916 paper on the shared-electron-pair chemical bond by G. N. Lewis and the several 1919 and 1920 papers by Irving Langmuir on this subject"* (From Zewail [6]). These theories were radically different from the first exposure to chemical bonds that he had as an undergraduate - the "Hooks and eyes" explanation to describe bonding between atoms. (According to this, there were a definite number of "hooks" on every atom that helped the atom to attach itself to another atom with a definite number of "eyes" that let other atoms to get attached to it).

Pauling took active part in student activities at the college. He was fearless in expressing his views in public – a trait which became more evident during his opposition to nuclear testing. It is said that on one occasion, during the dean's address to students, Pauling had no hesitation to stand up and correct some of the statements of the dean as he did not want the student community to be misinformed. In his final two years of undergraduate studies, Pauling decided to shift his focus to finding out the relation between the properties of substances and the structure of the atoms present in them. This opened up a completely new field of chemistry -quantum chemistry. By the end of the third year, Pauling had taken all the courses offered in physics and mathematics. As the college was short of faculty to teach chemistry, it requested Pauling who was still an undergraduate to teach a chemistry course for home economics majors. Pauling enjoyed this teaching experience. This chance assignment had a life changing impact on Pauling's personal life. Ava Helen Miller, his wife for 58 years, was one of the students taking the course.

After getting a B S degree in chemical engineering from Oregon Agricultural College in 1922, Pauling withstood pressure from his family to take up a job and decided to enroll for Ph.D in chemistry. He applied to both, University of California, Berkeley, and California Institute of Technology (Caltech), Pasadena. Pauling had to choose between the well established chemistry department at Berkeley where Lewis was the chairman and the new fledgling department at Caltech where A. A. Noyes was the chairman of the department. His dilemma was solved when he received the offer as a teaching fellow in chemistry with reasonable financial support from Caltech. He joined Caltech in 1922 and worked under Roscoe Dickinson on crystallography and with Richard Tolman on mathematical physics.

His experimental work involved using X-ray diffraction to determine the structures of selected crystals. Years later, Pauling considered his *"entry into the field of x-ray crystallography to be just about the most fortunate accident that I have experienced in my life"* (From Zewail [6]).

Pauling published several papers on the crystal structure of minerals during this period and was awarded the Ph.D degree in physical chemistry (with *summa cum laude*) with mathematics and physics as minors for his thesis titled "The Determination with X-rays of the Structure of Crystals" in 1925.

Noyes had been impressed by Pauling's extraordinary intellect during his stay at Caltech as a doctoral student and was keen that he should get post -doctoral experience in Europe to gain deeper understanding of quantum theory and the new advances being made by leading physicists to understand the structure of the atom. With this in mind, Noyes encouraged Pauling to apply for the newly established Guggenheim Fellowship immediately after obtaining his doctorate degree. Pauling submitted a proposal outlining his plans to utilize the fellowship to do postdoctoral studies in the new quantum physics in Europe and find ways to use principles of quantum physics to solve problems in chemistry. Pauling did not get the fellowship in 1925; instead he got a National Research Fellowship and he decided to utilize it to work under G. N. Lewis. As Noyes did not want to lose Pauling, he got Pauling financial help to remain in Caltech to continue his research on crystallography. This stay proved beneficial to Pauling as many leading physicists of Europe visited Caltech. Pauling got exposed to de Broglie's wave theory, Max Born's ideas of the structure of an atom as well as Heisenberg's work. Noyes was so certain of Pauling getting the Guggenheim Fellowship the following year that he not only arranged for Pauling to spend the year at various laboratories in Europe but also advanced $1500 for Pauling's initial expenses. Pauling was convinced about the importance of structure in understanding the nature of substances. He said "*It is structure that we look for whenever we try to understand anything. All science is built upon this search; we investigate how the cell is built of reticular material, cytoplasm, chromosomes; how crystals aggregate; how atoms are fastened together; how electrons constitute a chemical bond between atoms*" (From Pauling [7]).

Visit to Europe: 1920s is regarded as the golden period of physics as new theories and experiments to understand the nature and structure of the atom were revolutionizing physics. The accepted quantum theory was being questioned and new ideas and theories regarding atomic structure were emerging. Famous physicists - Niels Bohr, Werner Heisenberg, Wolfgang Pauli, Max Born and Erwin Schrödinger - working in different laboratories were contributing to new theories of atomic structure which replaced the old quantum theory with new quantum mechanics. Pauling left for Europe along with his wife in March 1926 and joined Sommerfeld's Institute for Theoretical Physics in Munich. Pauling also travelled to other leading laboratories in Europe and interacted with all the great physicists there culminating with a visit to Schrödinger's laboratory in Zurich. Pauling was granted an extension of the fellowship to work in Schrödinger's laboratory but was disappointed as he could not personally interact with Schrödinger. Schrödinger was too preoccupied with giving finishing touches to his wave theory to personally interact with this young post doctoral fellow from across the Atlantic. Pauling with his knowledge of mathematical physics, got the path breaking idea of combining Schrödinger's idea of wave mechanics with his own research findings in structures to understand how chemical bonds were formed. This exposure transformed Pauling's approach to chemistry and determined the future direction of Pauling's research. In early 1927, Pauling published one of his greatest papers in *Proceedings of the Royal Society (London)*, "The Theoretical Prediction of Physical Properties of Many - Electron Atoms and Ions. Mole Refraction, Diamagnetic Susceptibility and Extension in Space". As he had training in mathematical physics, Pauling was able to understand the principles of the theory of quantum mechanics. The most important thing Pauling brought back to the U S from Europe in 1927 was quantum mechanics! Pauling introduced the study of quantum mechanics without any delay at Caltech. He is rightly regarded as a pioneer of using quantum mechanics to understand the structure and function of atoms and molecules.

"*My year in Munich was very productive. I not only got a very good grasp of quantum mechanics — by attending Sommerfeld's lectures on the subject, as well as other lectures by him and other people in the University, and also by my own study of published papers — but in addition I was able to begin attacking many problems dealing with the nature of the chemical bond by applying quantum mechanics to these problems*" (From Zewail [6]).

Return to Caltech

During the period Pauling was in Europe, Robert Millikan as the president of Caltech had transformed the academic scenario at Caltech. The number of both undergraduate and graduate students had increased tremendously. Millikan and Noyes had put departments of physics and chemistry on the national map and Hunt Morgan, the noted geneticist was soon to join forces to start the department of biology. Pauling was appointed as Assistant Professor of Theoretical Chemistry at Caltech on his return. This fertile academic atmosphere suited his genius and he set a scorching pace working on a number of research projects as well as publishing his first book *The Structure of Line Spectra* with Samuel Goudsmit. Pauling soon decided that instead of fiddling around with new theoretical physics, he would concentrate on his first love - chemistry. He resumed work on crystal structure to understand molecular function but now his approach was different - he used X-Ray diffraction to get experimental data and used quantum mechanics to support the experimental findings with theory. *"Where the old quantum theory was in disagreement with the experiment, the new mechanics ran hand-in-hand with nature and where the old quantum theory was silent, the new mechanics spoke the truth"* (Linus Pauling, February 1929, From paulingblog.wordpress.com [8]).

Pauling had found a powerful tool in quantum mechanics to understand and solve not only the question of how and why chemical bonds were formed but also to comprehend more easily the laws that govern motion of bodies. Pauling's genius lay in his ability to combine the knowledge he had gained in Europe with his expertise in structure to find solutions to chemical problems. In his book "*Introduction to Quantum Mechanics with Applications to Chemistry*" (1935) with E. Bright Wilson, Pauling states "*the subject of quantum mechanics constitutes the most recent step in the very old search for the general laws governing the motion of matter*" (From paulingblog.wordpress.com [8]). Pauling applied principles of quantum mechanics to establish the correlation between molecular structure of atoms and compounds and their chemical properties and function.

Pauling visited Europe again in 1930 to interact with Lawrence Bragg in Manchester. Pauling was disappointed with his visit as he had no professional interaction with Bragg. However, he learnt the electron diffraction technique from Herman Mark to determine the structure of gas molecules. On his return to Caltech, he and his students built an apparatus to do electron diffraction of gaseous molecules and successfully solved the structures of a number of molecules.

Years of phenomenal work in structural chemistry

"In 1925...I had no way of distinguishing between the good ideas and the poor ideas about the electronic structure of molecules" - Linus Pauling (From Pauling [9]).

Pauling had proposed in 1928 that quantum mechanics could help find an answer to the question how chemical bonds between carbon atoms were formed but he had no computational support to prove his point. In 1930 physicist Slater working at MIT simplified Schrödinger's wave equation. By slightly modifying it, Pauling could arrive at the bond angle between carbon atoms and succeeded in calculating accurately the tetrahedral bond angle between carbon atoms. This work of Pauling was a game changer as it amalgamated structural chemistry with quantum mechanics and marked the birth of valence-bond theory - the final vindication of Lewis' theory of shared electrons and the chemical bond.

By 1931, Pauling was acknowledged as an authority on chemical bonds and quantum mechanics (quantum chemistry). Noyes, Pauling's mentor wrote in a letter to William Foster on October 15, 1931 saying *"I consider that the field of work in which Dr. Pauling is engaged, namely the study of the chemical bond and of valence from the standpoint of modern physics, is the most important line of research in theoretical chemistry today; and I venture to believe that there is no one in the world who in the same degree has chemical background and at the same time has the physical knowledge, mathematical power, and originality required for the handling of this problem"* (From scarc [10]).

Pauling received the first of the many awards that were showered on him when in 1931 he got the first Langmuir Prize awarded by The American Chemical Society (ACS) for the best young chemist of the US. The prize carrying a cheque for $1000 was given at the ACS meeting. Noyes called him *"the most promising young man with whom I have ever come in contact in my many years of teaching"* (From scarc [10]). Pauling was 30 years old then. In his response, Pauling said his greatest regret on that occasion was that his mother was not there to witness the award function.

Pauling Rules: Pauling combined his knowledge of quantum mechanics and experimental data from X-ray diffraction studies of molecules, combined with his understanding of structural chemistry and his emphasis on visualizing concepts through three dimensional models, arrived at a set of rules to predict the most likely molecular pattern that would occur in complex molecules. They were accepted by rcsearchers and came to be known as Pauling Rules. Pauling knew that chemists generally were reluctant to accept any theory with a lot of mathematics and accordingly he explained the basis of his rules through detailed examples of how his reasoning worked.

"*Pauling's model-building approach was novel to both crystallography and biological research. It became crucial to the investigations of protein structure, allowing precise visualization of the molecular arrangements and interactions hitherto hidden*" (From Kay [11]).

Pauling always thought of molecular structure in three dimensions. His models were built strictly according to all the known data (such as sizes of atoms and ions, bond lengths and angles obtained from X-ray crystallography). Building an exact model was a painstaking exercise. He would first draw the structure on paper and with his wife's help, fold the drawing at the appropriate angle and sew it to get a three dimensional model of the molecule's structure. This method enabled

Pauling to leapfrog to a complicated solution while lesser mortals got bogged down with the data from X-ray diffraction. Much later, Pauling successfully employed paper modeling to solve the structure of the alpha-helix of proteins.

The nature of the chemical bond

"I'd begun to think about the theory of the chemical bond very seriously in 1926, '27, after quantum mechanics was discovered and then in 1928 I published a paper, a preliminary paper, and said that I would write more later on. I didn't write anything more for three years because the problem turned out to be such a hard problem, the mathematical problem, that I couldn't solve it" (From Richter, *NOVA* /WGBH-Boston [12]).

Between April 1931 and June 1933, Pauling along with his coworkers published a series of papers on the nature of the chemical bond, the first one being *"The Nature of the Chemical Bond. Application of Results Obtained from the Quantum Mechanics and from a Theory of Paramagnetic Susceptibility to the Structure of Molecules"* published in *the Journal of the American Chemical Society* in 1931 (From Herman & Munro [13]). This was published in record time (a mere six weeks after the journal received the manuscript) as the editor felt there was no one competent enough to review the paper available! Pauling was aware of the importance of this paper and he says *"It seems to me that I have introduced into my work on the chemical bond a way of thinking that might not have been introduced by anyone else, at least not for quite a while. I suppose that the complex of ideas that I originated in the period of around 1928 to 1933……1931 was probably my most important paper — has had the greatest impact on chemistry"* (From Richter, *NOVA* /WGBH-Boston [12]). This paper formed the foundation upon which his magnum opus *"The Nature of the Chemical Bond"* was built. Pauling published six more papers between April 1931 and 1933.

Pauling's exhaustive study of the chemical bonds yielded three distinctive concepts - 1. hybridization of orbitals, 2. resonance in aromatic hydrocarbons and 3. Electronegativity scale. Robert J. Paradowski remarks *"Pauling's paper on bond energy and electronegativity proved to be highly influential. The qualitative concept of electronegativity as the ability of an atom in a molecule to attract electrons to itself was an old one. ....The importance of Pauling's paper derives from the fact that he was the first person to put this property on a numerical basis"* (From Paradowski [14]).

In 1937, Cornell University invited Pauling to deliver the George Fisher Baker lectures. Pauling was excited as it was the tradition at Cornell University to publish the Baker lectures as a monograph. Pauling had wished for a long time to compile all the work on the chemical bond in the form of a book by expanding content of the seven papers. Pauling knew that the resultant book would be the most comprehensive book available at that time on the subject of chemical bonds. As he did not want the monograph to be a mere reproduction of the lectures he delivered at the Cornell University, it took Pauling almost a year after his return to Caltech from Cornell to give the manuscript for publication. Cornell University published *"The Nature of the Chemical Bond and the Structure of Molecules and Crystals: An Introduction to Modern Structural Chemistry"* in 1939. It became a classic and remains one. Pauling had meant the book to be of use to senior graduate students of chemistry but its impact was felt across the scientific community.

Pauling presented chemistry as a natural consequence of the impact of quantum mechanics on the formation of chemical bonds. The wide appeal of the book was also due to its simple language. Pauling was justified in thinking that his work on the chemical bond "*probably has been most important in changing the activities of chemists all over the world - changing their ways of thinking and affecting the progress of the science*" (From Richter, *NOVA* /WGBH-Boston [12]). The book was responsible for changing the way scientists looked at chemistry and was an instant success getting rave response both for its content and its readability. It has inspired thousands of young students including one of the authors to take up chemistry as a career and to delve into interdisciplinary areas.

Pauling dedicated the book to G. N. Lewis, one of his academic heroes. Pauling wrote to Lewis on August 29, 1939 *"I am very happy to know that you are pleased with my book, and feel that it is good enough to be worthy of its dedication to you. You know, of course, that I had you in mind continually while it was being written, and I have been hoping that my treatment would prove to be acceptable to you"* (From scarc [15]). This book is considered as Pauling's greatest contribution to scientific literature. G. N. Lewis wrote to Pauling on August 25, 1939 *"I have just returned from a short vacation for which the only books I took were a half-dozen detective stories and your 'Chemical Bond'. I found yours the most exciting of the lot"* (From scarc [10]).

Pauling's *"Introduction to Quantum Mechanics, with Applications to Chemistry"* with E. Bright Wilson was the first book on quantum mechanics and was used by physicists and chemists alike during the early years to learn the subject. He then wrote the first publicly acclaimed book on General Chemistry for freshmen in chemistry.

Foray into biology

"By 1935......I felt that I had an essentially complete understanding of the nature of the chemical bond" - Linus Pauling (From Pauling [9]).

For more than a decade from the 1920s, Pauling had concentrated on understanding the nature of the chemical bond. By mid 1930s, Pauling's restless mind was looking for other areas to explore. He wanted to explore more complex organic molecules and his shift to biological molecules was probably motivated by the shift in the emphasis of funding by the Rockefeller Foundation from physical sciences to research in the area of interface between chemistry and biology. Pauling in his proposal for research grant to the Rockefeller Research Foundation wrote that his research on inorganic molecules would help in the study of organic substances such as proteins and other organic molecules. As he recounts later in 1933, he felt that "he *was fortunate in having a good understanding of two fields, structural chemistry and x-ray diffraction"* (From Pauling [16]).

Pauling chose hemoglobin as the biological molecule to focus on as he had always been fascinated by its chemistry. There were many reasons for this choice - first of all it was easily available and also like most scientists Pauling also believed that *"Proteins hold the key to the whole subject of the molecular basis of biological reactions"* (From Pauling [17]). He first concentrated on understanding the structure of hemoglobin and its magnetic properties and in 1935, published a paper on *"The Oxygen Equilibrium of Hemoglobin and Its Structural Interpretation"* with his co-worker Charles D. Coryell where they showed how the hemoglobin structure undergoes a change both when it gains or loses oxygen atom.

One of Pauling's great contributions to protein chemistry was his discovery of sicklc cell anemia as a molecular disease. When Pauling first heard of the change in the shape of blood cells in the venous blood of patients suffering from the disease sickle cell anemia (from Dr. William B Castle of Harvard Medical School). It occurred to Pauling *"Could it be that these patients can manufacture a special kind of hemoglobin such that the molecules are sticky and clamp on to one another to form long rods, which then line up side by side to form a long needle-like crystal, which as it grows inside of the red cell becomes longer than the diameter of the cell and thus twists the red cell out of shape?"* (From scarc [18]).

As predicted by Pauling, the red blood cells in the venous or deoxygenated blood of people with the disease sickled as they combined with themselves to form long rods and as they became longer than the diameter of the red cell, they changed their shape on being compressed. Pauling and coworkers showed how the abnormality of the cell structure inside the body caused the disease and argued that genetics determined the structure of proteins. Pauling strongly felt that his paper *"Sickle Cell Anemia, a Molecular Disease"* published in November 1949 in *Science* was responsible for starting the new multidisciplinary field of molecular biology comprising of biology, biochemistry, medicine and genetics. Francis Crick acknowledged that Pauling's work on sickle cell anemia was responsible to a large extent for merging genetics and protein chemistry. He further remarked *"I don't think it's right, really, to discuss the impact of Linus Pauling on molecular biology. Rather he was one of the founders of molecular biology. It wasn't that it existed in some way and he came down and put something on it. He was one of the founders who got the whole discipline going"* (From Crick [19]). Its impact was felt in the medical field also as it triggered interest in medical research in wide-ranging areas such as immunology, hematology of the human hemoglobin, pathology, serology as well as applied genetics.

The structure of the α-helix

While Pauling was Eastman professor at the University of Oxford in 1948, he decided to go back to the problem that had eluded a solution from as far back as the late 1930s- the problem of folding of amino acids in the polypeptide chain of α-keratin. When he was confined to bed with a bad cold, a bored Pauling decided to take the model building approach to arrive at the possible structure of a peptide chain. He drew a polypeptide chain of amino acids at approximately correct distances and bond angles on a plain sheet of paper. Then he folded the paper drawing at what he believed to be the correct angles. After many trials he succeeded in converting the linear drawing into a three dimensional model of a helical tube. Pauling discovered that this structure was common to the structure of ordinary things like hair, fingernails and animal horns and other fibrous proteins as well as to complex globular proteins like hemoglobin.

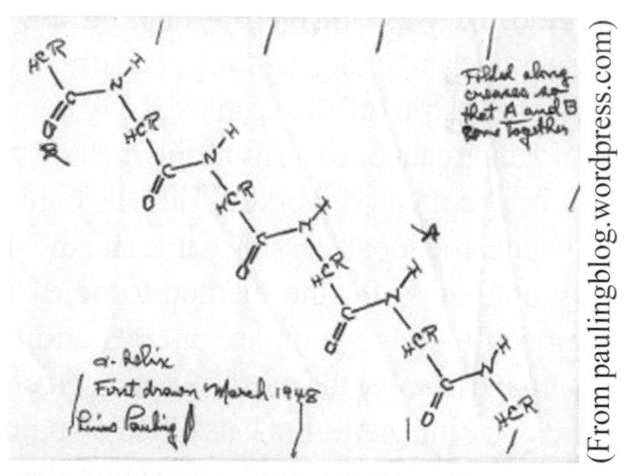

Reconstruction of the alpha-helix paper model. Drawn and folded by Linus Pauling

(From paulingblog.wordpress.com)

Pauling, Corey and Branson published the all important paper "The structure of proteins: Two hydrogen-bonded helical configurations of the polypeptide chain," in *Proceedings of the National Academy of Sciences* in 1951.

Jack Dunitz has said *"The formulation of the α-helix was the first and is still one of the greatest triumphs of speculative model building in molecular biology, and I am pleased that I helped to give it its name"* (From Dunitz [20]).

Max Perutz remarks *"When I saw the alpha-helix and saw what a beautiful, elegant structure it was, I was thunderstruck and was furious with myself for not having built this"* (From Perutz [21]).

"To divine the truth by conjecture"

Pauling had a unique method of arriving at amazing conclusions. His friend, Karl Darrow, called it the well documented stochastic (Greek word meaning "to divine the truth by conjecture") method which was in a way making an educated guess. This method could be applied to a hypothesis such as the structure of a molecule. When Pauling used the stochastic method, he was not merely making a wild guess. He used his vast knowledge and understanding of chemistry as well as of crystallography to carefully strip off all the unnecessary details until only one structure was left. Pauling cautioned on using this method without adequate understanding of how the method worked. He was convinced that this method is not another name for a guessing game and the investigator should be allowed only one guess for the method to be of use. He was able to use this approach to solve complex puzzles and this gift gave him the unique advantage to solve the most complicated chemical problems. It was not a mere guessing game, but was based on painstaking work and knowledge of chemistry. This method earned him the reputation of a magician who could come up with solutions where others had failed.

The first Nobel Prize in 1954

In a TV interview, Linus Pauling remarked *"In 1931 when my papers on the nature of the chemical bond appeared, Professor Noyes ....said that I probably would get the Nobel Prize someday. Well, I thought, that's nice of the old guy to say that, but I'm a little sceptical myself. And as the years went by, I thought, I don't do the sort of work for which Nobel Prizes are given"* (From Richter, *NOVA* /WGBH-Boston [12]).

Though Pauling had been tipped to win the Nobel Prize for his outstanding work on the nature of the chemical bond and in the field of proteins, it was late in coming. Year after year Pauling watched other lesser chemists being awarded the Nobel Prize. Pauling was not to be denied the prize as the great G. N. Lewis had been denied. Finally in 1954 came the announcement from Stockholm. As he was about to give a lecture at Cornell, Pauling first heard the news from a journalist who wanted to know his first reaction to the news (of being awarded the Nobel Prize). Pauling just wanted to know what the citation said and was delighted to learn that it was awarded *"For research into the nature of the chemical bond and its application to the elucidation of complex substances"* (From scarc [10]).

The faculty and students of Caltech hosted a dinner for the Paulings on December 3rd prior to their departure to Stockholm for the Nobel ceremony. It was a highly entertaining evening. The adoring faculty colleagues and students presented a two part skit "The Road to Stockholm" with rip roaring songs pulling Pauling's legs and funny sketches of Pauling.

"*Dr. Linus Pauling is the man for me / He makes violent changes in my chemistry / Oh, fie, when he rolls his eyes / All my atoms ionize*" (From scarc [22]).

Pauling was a popular Prize winner. A newspaper in Stockholm published a cartoon with a smiling Pauling at the center under the heading *"Prof. Linus Pauling stole the show at the Nobel banquet with his cheerful laugh"* with an accompanying funny poem titled "Theft" in Swedish for stealing the show.

Pauling addressed a large group of university students who had gathered to cheer him at Stockholm on December 10, 1954. He did not sermonize, but instead gave them the following message *"Perhaps as one of the older generation, I should preach a little sermon to you, but I do not propose to do so. I shall, instead, give you a word of advice about how to behave toward your elders, When an old and distinguished person speaks to you, Listen to him carefully and with respect — but do not believe him. Never put your trust in anything but your own intellect. Your elder, no matter whether he has gray hair or has lost his hair, no matter whether he is a Nobel Laureate, may be wrong. . . . So you must always be skeptical — always think for yourself"* (From scarc [10]). The students liked what they heard and wildly cheered Pauling. His message was splashed over all national dailies.

The pacifist

"It is sometimes said that science has nothing to do with morality. This is wrong. Science is the search for truth, the effort to understand the world; it involves the rejection of bias, of dogma, of revelation, but not the rejection of morality........... " (From Marinacci & Krishnamurthy [23]).

When America joined the Second World War, Pauling voluntarily joined war efforts and worked on a few military projects as he was against Nazism. Then the horror of Hiroshima and Nagasaki occurred. Ava Helen, a committed social activist and pacifist convinced Pauling that instead of trying to spread the antiwar message, he should actively lead the movement from the front to bring home the dangers of not only using atomic weapons to settle the outcome of wars but also testing atomic weapons in the atmosphere. Those were the days of furious arms race of the Cold War period and of witch hunting led by Senator McCarthy. Pauling was a target of this witch hunting. He was informed on February 14, 1952 that renewal of his passport was rejected as it was deemed that his travel abroad was detrimental to the interests of the US (From a letter from the passport office, February 14, 1952). Consequently he could not travel abroad even to attend scientific conferences. This prevented him from going to London for the conference where he could have seen the X-ray pattern of DNA recorded by Rosalind Franklin. His passport was given back only in 1954 so that he could attend the Nobel Ceremony. Most of his scientific colleagues deserted him as they felt that he was unwittingly acting as a spokesperson for communism. Even the American Chemical Society forsook him. Nothing deterred Pauling and he carried on courageously travelling the length and breadth of the US giving lectures highlighting the dangers of radioactive fallout. Ted and Ben Goertzel in the introduction to their book *Linus Pauling: A Life in Science and Politics* have highlighted the fact that Pauling *"used his scientific credentials to challenge the government's claims that fallout from nuclear testing was not harmful"*. He wrote a popular book "No More War" in 1958 and collected 10,000 or more signatures from prominent scientists and citizens from more than forty countries to present a memorandum to the UN to ban testing. His courage and total disregard for the personal price he had to pay ultimately won him a second Nobel Prize.

"*A Weird Insult from Norway*"[8] (LIFE magazine's headline)

Ironically, on October 10, 1963, the day the agreement for partially banning atmospheric testing of atomic weapons by the three nuclear powers (Russia, the US and the UK) was signed, the Nobel Peace committee announced the peace prize for 1962 to Pauling with this citation *"Linus Carl Pauling, who ever since 1946 has campaigned ceaselessly, not only against nuclear weapons tests, not only against the spread of these armaments, not only against their very use, but against all warfare as a means of solving international conflicts."* One of the members of the Prize committee felt that it was Pauling's relentless efforts that made the signing of the treaty to partially ban testing possible. Reporting the award of the prize, The New York Herald Tribune dated October 11, 1963 in an editorial titled "The Nobel Peacenik Prize," said that *"Award of the Nobel Peace Prize to Dr. Pauling, whatever the committee's reasons, inevitably associates this semi sacrosanct honor with the extravagant posturings of a placarding peacenik."* In contrast to the joyous reaction to his first Nobel Prize, Caltech was antagonistic and cold shouldered him. Pauling felt so unwanted that he resigned from Caltech in 1964 with a heavy heart – the institution he had entered as a graduate student and served with distinction for four decades. He wandered from one academic institution to another (University of California, Santa Barbara, and San Diego,

Stanford University). He remained professionally active and a committed activist for social justice and peace till the end.

Orthomolecular medicine and vitamin C

While at Caltech Pauling had conducted studies in the 1930s and 40s on whether certain psychiatric disorders were the result of some genetic and biochemical malfunctioning. He was convinced that mental illnesses could be treated with better results if the patient was administered vitamins, particularly Vitamin C, in sufficiently large doses. He was convinced about the importance of vitamin C both as a palliative and prophylactic if taken in mega doses (particularly for combating common cold). He became almost fanatical in his advocacy of vitamin C as an antidote and a cure for a variety of diseases ranging from cancer, cardiovascular problems, neurological disorders like schizophrenia, aging related degeneration and infections. The strong public statement that mega doses of vitamin C was the panacea for every leaking nose made by so pre-eminent a scientist as Linus Pauling *"caused such a run on vitamin C that pharmacies were unable to maintain adequate supplies, and .....seldom has a pronouncement caused such a run on anything"* (From Wall Street Journal [24]).

Pauling The Teacher

Pauling was an inspirational teacher. He believed that budding chemists were attracted not by theoretical aspects of chemistry but by chemical reactions and that chemical education should create a sense of wonder for the subject in students. To create this sense of wonder, he made generous use of tricks (as magicians do) and interesting experiments to show rather than tell the students how the subject worked. He used drawings of the structure of complex molecules and three dimensional molecular models to illustrate complicated concepts. Just like Feynman in physics, he enjoyed teaching the freshman class. One of the students describing his impact on students recalled that Pauling was *"wholly informal in dress and appearance. He bounded into the room, already crowded with students eager to see the Great Man, spread himself over the seminar table next to the blackboard and, running his hand through an unruly shock of hair, gestured to the students to come closer. . . The talk started with Pauling leaping off the table and rapidly writing a list of five topics on which he could speak singly or all together. He described each in a few pithy sentences, including racy impressions of the workers involved"* (From scarc [25]). His lectures were spectacular and often dramatic. He was a master at creating an effect. Professor Dunitz describes one such dramatic effect *"A large beaker filled with what looked like water stood on the bench. Pauling entered, picked a cube of sodium metal from a bottle, tossed it from hand to hand (done safely if your hands are dry) and warned of its violently explosive reaction with water. He then threw it into the beaker. As students cowered in fear of an explosion, he said nonchalantly "but its reaction with alcohol is much milder"* (From Perutz [26]).

There are many anecdotes about Pauling as a teacher. He knew that students were in awe of him and revered him almost as a god. After teaching freshman chemistry for many decades, Pauling's associate Jürg Waser took over the responsibility. Tom Hager in his book "Force of Nature" narrates one funny anecdote. According to him, one day Pauling came as a guest lecturer to the class and as hc cntered the class, he heard students laughing loudly. Someone had written on the blackboard "Pauling is God and Waser is his prophet." Pauling, wanting to know the cause for the merriment, turned back, looked at the blackboard, coolly wiped out "Waser is his prophet," and delivered his lecture to the class!

In a Letter to A. A. Noyes written on November 18, 1930, Pauling wrote *"...To awaken an interest in chemistry in students we mustn't make the courses consist entirely of explanations, forgetting to mention what there is to be explained"* (From scarc [10]).

What was crucial to the vast number of young students who came under his spell as a teacher was Pauling's vision of the content of chemistry. He wanted chemistry to be taught as an unified discipline where experimental evidence was backed by theoretical proof. His classes reflected this new approach and based on his lecture notes, he published in 1946 the book on General Chemistry.

Personal life

"I have been especially fortunate for about 50 years in having two memory banks available-whenever I can't remember something I ask my wife, and thus I am able to draw on this auxiliary memory bank. Moreover, there is a second way In which I get ideas ... I listen carefully to what my wife says, and in this way I often get a good idea. I recommend to ...young people .... that you make a permanent acquisition of an auxiliary memory bank that you can become familiar with and draw upon throughout your lives" (From Goertzel & Goertzel [27]).

Pauling married Ava Helen Miller, his former student from Agricultural College, Portland, in 1923 and shared a very close personal relation with her for the next six decades. He completely depended on her for intellectual, academic as well as moral support and above all for total companionship that lasted till her death in 1981. In his acceptance speech at the Nobel Peace Prize Award function Pauling called her, his "*constant and courageous companion and coworker.*" He could never come to terms with her death. He said in an interview that he gave in 1990 to the Academy of Achievement at Big Sur California, *"since my wife died...I don't have anything to do now, except make discoveries and write papers."* She was a committed peace activist and she influenced Linus to take up antinuclear testing at great personal cost. In the interview he gave to NOVA of PBS TV channel in June 1977, Pauling said of his wife *"The humanistic concern she had was very great. I'm sure that if I had not married her, I would not have had this aspect of my career — working for world peace"* (From scarc [28]).

A brilliant achiever and a fearless crusader

Pauling made outstanding contributions to many fields by his original ideas and out of the box thinking. He was a pioneer in quantum chemistry, structural chemistry, molecular biology and medicine. His contributions were even more extraordinary as they were in the overlapping areas of these disciplines. In an interview given at Big Sur, California on 11 November 1990, Pauling responded to the often asked question "How does it happen that you have made so many discoveries? Are you smarter than other scientists?" *"And my answer has been that I am sure that I am not smarter than other scientists. ...... I think ... harder, think more than other people do, than other scientists. .....almost all of my thinking was about science and scientific problems that I was interested in"* (From achievement.org [29]).

This single minded pursuit of science is reflected in the lecture "Chemical Achievement and Hope for the Future." that he gave at Yale University, in October 1947. He told the audience *"Science cannot be stopped. Man will gather knowledge no matter what the consequences – and we cannot predict what they will be. Science will go on ....I know that great, interesting, and valuable discoveries can be made and will be made... But I know also that still more interesting discoveries will be made that I have not the imagination to describe — and I am awaiting them, full of curiosity and enthusiasm"* (From Baitsell [30]).

Pauling was convinced if anyone wants to *"make discoveries, it's a good thing to have good ideas. And second, you have to have a sort of sixth sense — the result of judgment and experience — which ideas are worth following up. I seem to have the first thing, a lot of ideas, and I also seem to have good judgment as to which are the bad ideas that I should just ignore, and the good ones, that I'd better follow up"* (From Pauling & Kamb [31]).

When asked how he could come up with so many good ideas he apparently said *"If you want to have good ideas you must have many ideas. Most of them will be wrong, and what you have to learn is which ones to throw away"* (From Crick [19], Francis Crick quoting Pauling in "*The_Impact of Linus Pauling on Molecular Biology*"). Pauling was not shy of admitting that his idea could be wrong.

Francis Crick has said *"Pauling was not always right in his ideas. But my belief is that, in most cases, if somebody is always right in his ideas you find that he does not have much to say. It is an expression of somebody's fertility that he does produce quite a number of ideas, and I think Linus Pauling's score is pretty high..."* (From Crick [19]).

Max Ferdinand Perutz saw *"[Linus Pauling] as a brilliant lecturer and a man with a fantastic memory, and a great, great showman. I think he was the century's greatest chemist. No doubt about it"* (From scarc [32]).

Sten Samson in an Interview by Anthony Serafini for *Linus Pauling: A Man and His Science* (1984), paid this tribute *"[Pauling] looks at the forest and lets other people...work out the specific individual things in detail; he has a terrifically lively intellect, reading [Pauling's] paper, the information here is just tremendous, the ideas flow out of the pen, and there are several lifetimes of work...to be done"* (From scarc [10]). This sums up Pauling's genius for generating path-breaking ideas in multiple fields.

Pauling was a flamboyant showman in everything he did - be it teaching undergraduate chemistry, announcing a scientific discovery or protesting outside the White House. He was not afraid of taking unpopular stands where necessary and was a fearless crusader fighting powerful opponents including the President of the United States! When Kennedy announced resumption of atmospheric tests, Pauling sent a telegram to the President calling the decision immoral. Yet when President Kennedy decided to honour all the American Nobel winners with an evening of music followed by a banquet on 29th April, 1962, Paulings were very happy to be among the invitees. A day before that big event, Pauling, carrying a "Stop Testing Now" placard, was among the three thousand people picketing outside the White House protesting against the Presidential decision. He joined the group again next morning but changed into formal dress in the evening and attended the Presidential banquet! It was a joyous evening with President and Mrs. Kennedy being gracious hosts. Reporters covering the occasion noticed Mrs. Kennedy in serious conversation with Pauling and asked Pauling later what was the conversation about, Pauling replied "*Mrs. Kennedy said, 'Dr. Pauling do you think that it is right to march back and forth out there in front of the White House carrying a sign and cause Caroline to say, 'Mummy, what has Daddy done wrong now?' I thought that was pretty clever"* (From scarc [28]).

Pauling had strong work ethics inherited from his father. Answering a student's question about his working ethics, Pauling told the audience that *"I always follow my golden Rule. It goes something like this: "Do unto others twenty-five percent better than you expect them to do unto you." ... The twenty-five percent is for error"* (From WikiQuote [33]).

Pauling is the only scientist with two full Nobel Prizes. He probably would have got another Nobel Prize for the DNA structure if he could have seen Rosalind Franklin's X-ray photographs in London in 1952.

Final years

Pauling was a committed scientist and worked tirelessly till the very end. He worked on nuclear structure and other topics, though not with much success. When he passed away in 1994, New York Times led eulogies with an obituary tracing the extraordinary journey of not only the greatest chemist but also the committed humanist.

"Linus C. Pauling, a brilliant chemist and an untiring political activist who received one Nobel Prize for chemistry and another for peace, died on Friday at his ranch in the Big Sur area of Northern California. He was 93" (From New York Times [34]).

"The world lost one of its greatest scientists and humanitarians and a much respected and beloved defender of civil liberties and health issues. Because of his dynamic personality and his many accomplishments in widely diverse fields, it is hard to define Linus Pauling adequately. A remarkable man who insistently addressed certain crucial human problems while pursuing an amazing array of scientific interests, Dr. Pauling was almost as well known to the American public as he was to the world's scientific community" (From Marinacci [1]).

Pauling was a colossus striding the world stage and the world will have to wait a long while to see another scientist like him.

References

1. Barbara Marinacci, *"Linus Pauling — In Memoriam"*, Linus Pauling Institute of Science and Medicine, 1994.
2. Clifford Mead and Thomas Hager (Ed.) *"Linus Pauling: Scientist and Peacemaker"*, (Chapter one - Roots of Genius (Tom Hager)). Corvallis: Oregon State University Press, 2001.
3. Thomas Hager, *Force of Nature: The Life of Linus Pauling*. New York: Simon & Schuster, 1995.
4. Robert J. Paradowski: Typescripts (*Snapshots of Pauling's Childhood in Condon*, posted March 3, 2009, scarc in The Pauling Blog Website).
5. Barbara Marinacci (Ed.), Introduction by Linus Pauling, *Linus Pauling In His Own Words: Selections from his writings, speechs, and interviews,* Simon & Schuster, New York (1995).
6. Ahmed H. Zewail, *"The Chemical Bond: Structure and Dynamics"*, Academic Press, ed. 1992.
7. Linus Pauling, The Place of Chemistry In the Integration of the Sciences', *Main Currents in Modern Thought* (1950), **7**, 110.
8. From paulingblog.wordpress.com, The Oregon State University Libraries Special Collections & Archives Research Center (scarc).
9. Linus Pauling, "Fifty years of progress in structural chemistry and molecular biology," *Daedalus*, 99 (Fall 1970): 988-1014. 1970 (From 'Linus Pauling The Nature of the Chemical Bond' – A documentary History, The Oregon State University Libraries SCARC).
10. 'Linus Pauling, The Nature of the Chemical Bond' – A documentary History, The Oregon State University Libraries Special Collections & Archives Research Center (scarc).
11. Lily E. Kay, *The Molecular Vision of Life: Caltech, The Rockefeller Foundation and the Rise of the New Biology*, New York: Oxford University Press 1993. (From Linus Pauling & the structure of proteins – A documentary History, Special Collections & Archives Research Center, The Valley Library, Oregon State University, scarc).

12. Robert Richter, *"Linus Pauling, Crusading Scientist."* 1977 produced for *NOVA* /WGBH-Boston (From 'Linus Pauling The Nature of the Chemical Bond' – A documentary History, The Oregon State University Libraries Special Collections & Archives Research Center, scarc).
13. Z. S. Herman and D. B. Munro, (J. Am. Chem. Soc. 53, 1367-1400 (1931); From *The publications of Professor Linus Pauling*, The Linus Pauling Institute of Science and Medicine.
14. R. J. Paradowski, *The Structural Chemistry of Linus Pauling*, pg. 450. 1972. (From 'Linus Pauling - The nature of the chemical bond – A documentary History', The Oregon State University Libraries Special Collections & Archives Research Center, scarc).
15. Letter from Linus Pauling to G. N. Lewis. August 29, 1939. Page 1 (From Linus Pauling Day-by-Day (Webpage) Special Collections & Archives Research Center, Oregon State University, scarc).
16. Linus Pauling, *"How My Interest in Proteins Developed."* January 12, 1993. (From 'Linus Pauling & the structure of proteins' – A documentary History, Special Collections & Archives Research Center, The Valley Library, Oregon State University, scarc).
17. Linus Pauling. "Signs of Life." *Electronic Medical Digest*, 35-36. 1949 (From 'Linus Pauling & the structure of proteins' – A documentary History, Special Collections & Archives Research Center, The Valley Library, Oregon State University, scarc).
18. *"Interview with Linus Pauling."* 1960 Produced by the National Film Board of Canada (From 'Linus Pauling & the structure of proteins' – A documentary History, Special Collections & Archives Research Center, The Valley Library, Oregon State University, scarc).
19. Francis Crick *"The Impact of Linus Pauling on Molecular Biology"* February 28, 1995. (From 'Linus Pauling & the structure of proteins' – A documentary History, Special Collections & Archives Research Center, The Valley Library, Oregon State University, scarc).
20. Jack Dunitz. *"La Primavera."* 2011, (From 'Linus Pauling & the structure of proteins' – A documentary History, Special Collections & Archives Research Center, The Valley Library, Oregon State University, scarc).
21. Max Perutz in BBC programme, 'Lifestory: Linus Pauling' (1997). (From 'Linus Pauling & the structure of proteins' – A documentary

History, Special Collections & Archives Research Center, The Valley Library, Oregon State University, scarc).

22. Song lyrics from *"The Road to Stockholm."* 1954, Chemistry-Biology Stock Company, C.I.T (From Linus Pauling - The nature of the chemical bond – A documentary History, The Oregon State University Libraries Special Collections & Archives Research Center, scarc).
23. Barbara Marinacci and Ramesh Krishnamurthy (Eds), "*Linus Pauling on Peace: A Scientist Speaks Out on Humanism and World Survival*", 1998.
24. From The Wall Street Journal report 1970 quoted by New York Times dated 21 August, 1994 in its Obituary of Pauling.
25. From Linus Pauling The Nature of the Chemical Bond - A documentary History, Narrative 43, The Teacher, Special Collections & Archives Research Center, Oregon State University Libraries, (scarc.library.oregonstate.edu).
26. M. Perutz, *Linus Pauling,* Nature Structural Biology, p2.
27. Ted George Goertzel and B. Goertzel, *Linus Pauling: A Life in Science and Politics*. Basic Books, 1995.
28. 'Linus Pauling And the international Peace Movement'- A documentary History - Special Collections & Archives Research Center, Oregon State University, scarc).
29. From Linus Pauling Interview, Academy of Achievement, February 29, 2008, Retrieved Jan13, 2015 from http://www.achievement.org/autodoc/page/pau0int-3
30. George A. Baitsell, ed., Reproduced in *Science in Progress*. Sixth Series. 100-121, 1949. October 15, 1947 (From A documentary History of Linus Pauling, Hemoglobin, Sickle cell Anemia, Special Collections & Archives Research Center, Oregon State University, scarc).
31. (Quoted by Nancy Rouchette, *The Journal of NIH Research* (Jul 1990), **2**, 63) Linus Pauling, Barclay Kamb, *Linus Pauling: Selected Scientific Papers*, Volume II, World Scientific (2001).
32. From transcript of audio of Max Perutz in BBC programme, 'Lifestory: Linus Pauling' (1997). On 'Linus Pauling and the Race for DNA' –- A documentary History, Special Collections & Archives Research Center, Oregon State University (scarc).

33. Pauling's reply to an audience question following his lecture at Monterey Peninsula College, in Monterey, California, 1961. (From WikiQuote – Linus Pauling).
34. New York Times obituary – *Linus C. Pauling dies at 93, Chemist and Voice for Peace* dated August 21, 1994.

"Linus Pauling - Biographical". Nobelprize.org. Nobel Media AB 2014. Web. 15 May 2015. <http://www.nobelprize.org/nobel_prizes/chemistry/laureates/1954/pauling-bio.html>

https://paulingblog.wordpress.com/tag/alpha-helix/, The Oregon State University Libraries Special Collections & Archives Research Center.

"Robert S. Mulliken - Biographical". Nobelprize.org. Nobel Media AB 2014. Web. 18 May 2015. <http://www.nobelprize.org/nobel_prizes/chemistry/laureates/1966/mulliken-bio.html>

Molecular Orbital Theory

The valence bond method was propounded by Linus Pauling. An alternative, powerful method to understand chemical bonding and electronic structure of molecules is by the use of molecular orbitals. The method was developed by Mulliken in collaboration with Hund and later championed by Mulliken. The method has been employed very widely and has proved to be most effective in the study of molecular structure and spectroscopy. Pauling was lukewarm about molecular orbital theory.

R. S. Mulliken (born on June 7, 1896 in Newburyport, Massachusetts in the US) obtained his Ph. D degree from Chicago University where he worked as a professor of chemistry and physics for many years. He received the Nobel Prize for chemistry in 1966.

(From Nobelprize.org)

20. ROBERT BURNS WOODWARD (1917–1979)

Artist in organic synthesis

"Woodward is to organic chemistry what Louis (Armstrong) is to the trumpet!" At that, my friend turned around slowly, looked Bob in the eye, and said, "Man, you must be one hell of a chemist!" Bob said he thought that was the most sincere compliment he ever got (From Milner [1]).

(From Nobelprize.org)

Early years

Robert Woodward was born on April 10, 1917 in Boston, Massachusetts, USA, as the only child of Arthur Chester Woodward and Margaret Burns. His father died in the influenza epidemic of 1918 when Robert was just a year old. He had a difficult childhood as his mother who remarried was abandoned by her husband shortly after the marriage. His mother had to raise her young son under financially difficult circumstances. His education up to high school was in public schools in Quincy, Massachusetts. During this period, Robert spent much of his time with his grandfather, Harlow Elliot Woodward, who was an apothecary. Many feel that Woodward's interest in chemistry was perhaps inspired by the time he spent at his grandfather's shop.

Woodward showed interest in chemistry even as a child. When he was eight years old, he got an elementary chemistry kit as a gift. This sparked his lifelong passion for chemistry. He started conducting experiments on his own using the kit and got even more interested in experimental chemistry. He enlarged the scope of this kit by adding experiments from undergraduate textbooks. He bought Ludwig Gattermann's *Practical Methods of Organic Chemistry,* the popular book used to teach experimental organic chemistry at college level and successfully conducted most of the experiments in the book even before he started high school. He showed great enterprise for one so young (he was only eleven years old) and obtained copies of a few original research papers published in German journals through the Consul General of the German consulate in Boston. One of the papers he obtained was the path breaking paper of Diels and Alder, detailing the reaction named after them (the Diels-Alder reaction). Young Woodward was so fascinated by this reaction that he used it repeatedly in his experimental as well as theoretical research throughout his research career. His creative imagination and ability to visualize were apparent even when he was a boy. Later he would recall that as a boy he had his own benzene formula and that he spent a lot of time trying to dream of synthesizing his own benzene structure.

Diels–Alder reaction

In 1928, Otto Paul Hermann Diels and Kurt Alder of Germany were the first to successfully describe the cycloaddition reaction between a dienophile or a substituted alkene and a conjugated diene to yield a substituted cyclohexene. They found that the product of the chemical reaction formed a six-membered ring of carbon atoms in which only one pair of carbon atoms was bound together by a double bond instead of extending in a linear fashion. The reaction was named after them.

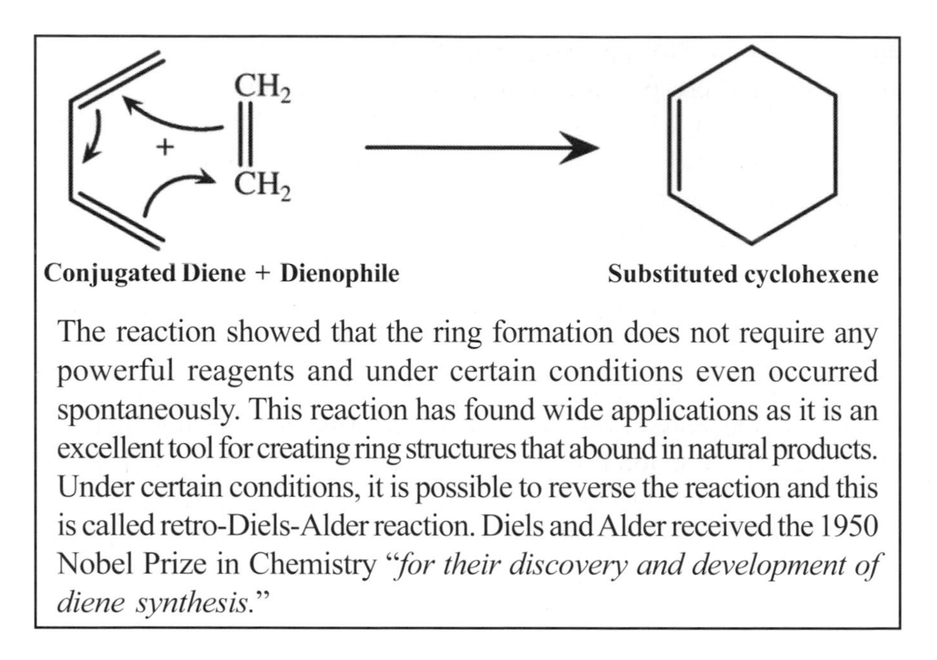

The reaction showed that the ring formation does not require any powerful reagents and under certain conditions even occurred spontaneously. This reaction has found wide applications as it is an excellent tool for creating ring structures that abound in natural products. Under certain conditions, it is possible to reverse the reaction and this is called retro-Diels-Alder reaction. Diels and Alder received the 1950 Nobel Prize in Chemistry "*for their discovery and development of diene synthesis.*"

Woodward joined MIT in 1933 when he was 16 years old as an undergraduate in the chemistry department. He soon realized that his knowledge of organic chemistry was far above that taught in the freshman course. After the initial excitement as a freshman (when he made an impression on Professor James Flack Norris who turned out to be his guardian angel), he found the formal course requirements of the undergraduate course too boring and lost interest in completing his assignments. He did not attend chemistry classes regularly. He neglected his studies so badly that he was expelled from the undergraduate programme at the end of the fall term of the second year (sophomore year) in 1934 for neglecting course requirements. On dropping out of the programme, he joined the food technology department of MIT as an employee. Fortunately for the future of organic chemistry, Professor Norris who had recognized Woodward's unique talent did not wish to lose him. He traced Woodward and persuaded him to rejoin the chemistry department and complete the course by giving a written exam instead of completing the course work of the sophomore year. After obtaining apologies for his cavalier attitude of the sophomore year and promise of fulfilling the requirements of the course, MIT readmitted Woodward in the fall term of 1935. There was no looking back this time around. He became one of the hardest working and most focused student. Woodward completed the B S degree next year and received the doctorate degree just a year later in 1937.

Professor James F Norris remarked *"We saw we had a person who possessed a very unusual mind and we wanted to let it function at its best...... We did for Woodward what we have done for no other person like him in our department. We think he will make a name for himself in the scientific world"* (From Norris [2]). Woodward certainly fulfilled this prophecy.

After obtaining his doctorate degree, Woodward joined the University of Illinois as a postdoctoral fellow. After a few months, he joined Harvard University as a Junior Fellow. He advanced rapidly in the academic hierarchy becoming a full professor in 1950 at the age of 33 years (Morris Loeb Professor of Chemistry three years later, and Donner Professor of Science in 1960).

Early contributions

Woodward was one of the earliest chemists to apply instrumental techniques that had just become popular to arrive at the structure of complex molecules and natural products. For example, he used ultraviolet spectroscopy to unravel the structures of complex molecules strychnine and terramycin among other organic molecules. The manner in which Woodward arrived at solving the structures of terramycin and strychnine are classic examples of his genius. Derek Barton spent a year at Harvard when Woodward himself was on sabbatical (at Harvard itself) and had a close view of Woodward's genius at work. The terramycin molecule was industrially

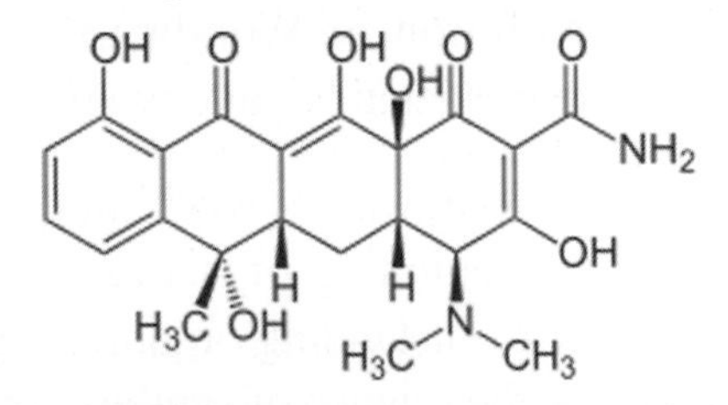

important and many chemists were involved in solving its structure. There was a plethora of data which while experimentally correct, was confusing. While less gifted organic chemists were struggling to solve the puzzle, according to Barton, Woodward systematically wrote all the known facts on a large cardboard piece and solved the structure of terramycin by interpreting the assembled data. Barton remarked that no other chemist could have solved the structure of terramycin then.

Based on the vast empirical data that he had collected in the course of his own work and data gathered from others work, Woodward proposed a set of rules in four papers published between 1941 and 1942. These rules were further refined by Louis and Mary Fieser. Woodward Rules are sometimes referred to as Woodward-Fieser rules. Woodward proposed that they could be used to arrive at the structures of newly discovered natural products as well as of molecules synthesized in the laboratory. These rules illustrate Woodward's penchant for arranging scientific data in an orderly and meticulous manner and a gift for analysis of raw data.

"The structure known, but not yet accessible by synthesis is to the chemist what the unclimbed mountain, the uncharted sea, the untilled field, the unreached planet are to other men..... The unique challenge which chemical synthesis provides for the creative imagination and the skilled hand ensures that it will endure as long as men write books, paint pictures, and fashion things which are beautiful, or practical, or both"- RBW (From Bowden & Benfey [3]).

Quinine, ferrocene and strychnine

These three molecules tell their own stories in the evolution of Woodward as the undisputed leader in organic synthesis.

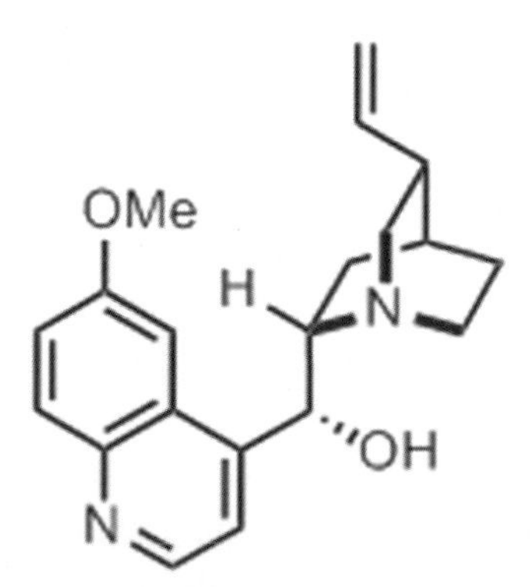

Quinine: Woodward's work on quinine marks the beginning of his remarkable career as an outstanding organic chemist who made the science of synthesis an art. This work was directly related to World War II efforts. In the early 1940s, Japanese occupied parts of Southeast Asia rich in cinchona plantations, the only source of quinine, the ideal anti-malarial drug. Woodward was aware of the romance associated with quinine's unique medicinal property. Quechua

Indians of Peru are credited with discovering quinine in the bark of cinchona tree - a tree native to Peru. According to the popular story, the bark had been used to successfully treat the wife of the Viceroy of Peru who was suffering from malaria. There was an urgent need to chemically synthesize quinine. At this time, synthesis was not yet systematized and organic chemists believed that complex molecules could not be synthesized in the laboratory. Woodward and his post-doctoral assistant, William von Eggers Doering were successful in synthesizing the alkaloid quinine in 1944 after four years of intense work. This was a major breakthrough in organic synthesis. What aided Woodward in successfully synthesizing quinine was his realization that Paul Rabe of Germany had succeeded as far back as in 1905 in converting quinotoxine to quinine. Woodward followed the same route to synthesize quinine. Even though it was a long and tedious route for the bulk production of the chemical, it aided other chemists to successfully develop other anti-malarial drugs such as chloroquine, primaquine, and mefloquine. Woodward was only 27 years old at that time and this synthesis established his credentials as a unique synthetic organic chemist. Woodward and his coworkers succeeded in synthesizing cholesterol and chlorophyll. These involved elegant reaction sequences unknown at that time. Turning his attention to antibiotics, Woodward and co-workers produced tetracycline in 1962 and cephalosporin C in 1965.

Ferrocene: This is illustrative of Woodward's unique ability to intuitively go to the core of problems outside his own research interests. Woodward's dual interest in aromatic compounds and the Diels-Alder reaction made him take interest in the newly created organometallic sandwich compound where an iron atom was sandwiched between two parallel cyclopentadienyl rings. Even though Woodward was not working on organometallic compounds, he succeeded in proposing the correct structure for the compound which was later named ferrocene.

Strychnine: Woodward in a lecture at Zurich in 1954 said *"Strychnine has been so artfully constructed that the task of wresting from Nature the secret of its architecture assumed classical stature"* (From Benfey & Morris [4]).

Strychnine was known for its toxic properties even in the 19th century. It was obtained from the button shaped hard seeds of a tropical tree and was used to kill rats. However, solving the structure of strychnine had proved to be elusive. Hendrickson of Brandeis University had called strychnine the Mount Everest of structural organic chemistry. (From Benfey & Morris [4]) Many including Robert Robinson of England had been successful in only partially solving it. Woodward was so fascinated by its complexity that he took up the challenge. He built on the results of the works of others, particularly of Leuchs and Robert Robinson and arrived at the complete structure with seven fused rings for the first time. At the time of announcing this path breaking discovery in 1948, he acknowledged the contributions of Leuchs *"for his beautiful experimental work"* and of Robinson who according to Woodward *"painted virtually the whole picture"* (From Benfey, Morris [4] & Woodward [5]).

Robinson, Woodward and strychnine

"It was very difficult for Sir Robert to realize that another intelligence, fully the equal of his own, had entered upon the alkaloidal scene. R. B. Woodward liked to tell the following story about the structure of strychnine. In 1947, Woodward and Robinson had dinner together in a New York restaurant, Sir Robert being the host. As the meal progressed, Woodward told Robinson that he had been thinking about the structure of strychnine. Since Robinson had a group of fifteen or so workers concentrated on this subject, he had to admit that the problem was interesting. Woodward asked Robinson what he thought about the true — as later revealed — structure of strychnine and wrote it on the paper tablecloth. Robinson looked at it for a while and cried in great excitement, "That's rubbish, absolute rubbish!" So ever after, Woodward called it the rubbish formula and was indeed quite surprised to see, a year later in Nature, that this was also the formula deduced eventually by Robinson. Woodward should have sent the tablecloth for publication!" (From Barton [6])

After the synthesis of strychnine in 1954, Woodward was treated like an emperor of organic chemistry.

Vitamin B_{12} (Woodward's magnum opus): Synthesis of vitamin B_{12} is the greatest testimony to Woodward's genius and to collaborative research. Woodward and Albert Eschenmoser of the Federal Institute of Technology, Zürich, decided to collaborate in this mammoth task. It was decided that each group would concentrate on one half of the problem. The ETH group headed by Eschenmoser started its work on the synthesis of B_{12} molecule in 1960 and the Harvard group headed by Woodward in 1961. Woodward was entrusted with the task of putting the two halves together to achieve the total synthesis. While the Harvard group took the thermal approach to arrive at the pathway, ETH group took the photo-chemical route. The joint project involving more than one hundred PhD students and postdoctoral fellows and others for nearly 12 years resulted in the synthesis of this complicated coenzyme involving over 100 sequential steps. Woodward announced in 1972 *The total synthesis of vitamin B_{12}* in his lecture given in the IUPAC international conference held in New Delhi (India) in July. The spin-off of this synthesis was the formulation of the theory of orbital symmetry conservation with Roald Hoffmann.

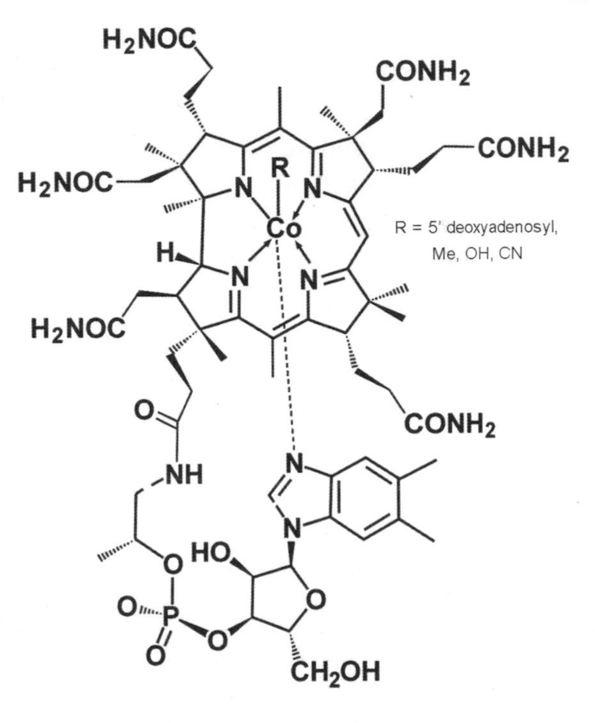

Woodward – Hoffmann Rules
Woodward had noticed during the synthesis of vitamin B_{12} that a crucial reaction did not proceed as predicted by him. He had noticed previously

some reactions proceeding in a totally unanticipated manner. He arrived at the reason for this anomaly. He requested his younger colleague Roald Hoffmann at Harvard to verify theoretically his experimental findings. After four years of work, they were able to arrive at a set of rules to explain the unpredictability of the nature of the products in some reactions when the same two compounds yielded different end products when the reactions were activated by different methods. The rules came to be known as Woodward-Hoffmann rules. These rules enable chemists to predict with a fair degree of certainty the products of reactions between two compounds, based on whether the reaction is thermally activated or photochemically activated. Hoffmann received the Nobel Prize in 1981 along with Fukui.

Master of synthesis

Woodward synthesized a mindboggling number of complex molecules of natural products, steroids, antibiotics and proteins. 1944 – quinine, 1945-1947 – chemical structure of penicillin, 1948 – patulin, 1947-1949 – strychnine (structure), 1951 – Cholesterol, cortisone, 1952 – ferrocene, 1954 – cevine, 1958 – gliotoxin, 1959 – ellipticine, 1954 – lysergic acid (basis of LSD) and strychnine (synthesis), 1956 – reserpine, (tranquillizing drug), 1960 –calycanthine, oleandomycin and chlorophyll, 1962 – tetracycline, 1963 – streptonigrin and colchicine, 1964 – tetrodotoxin and 1965 – cephalosporin C and 1972 – his crowning achievement vitamin B_{12}. This extraordinary period is often referred to as the Woodwardian era during which he successfully demonstrated that it was possible to synthesize complex natural products by meticulously planning and carefully applying the principles of chemistry. It was perhaps for the first time that natural product synthesis was not only utilitarian but also elegant.

Woodward is credited with making organic synthesis an art. Woodward's unique gift was not so much in creating new methods of synthesis as in his ability to collate all the available data to solve intricate puzzles of the structure of organic molecules. This gift combined with his mental discipline and ability to organize systematically all available data enabled him to visualise at once the problem in its entirety and arrive at the solution. The hallmark of his approach to the synthesis of complex molecules was the planning of every step backed by chemical intuition. He became famous for his innovative approach wherein intellectual discipline was a prerequisite for successful synthesis of complex molecules.

Woodward pioneered the method of 'stereoselective synthesis', or synthesizing compounds having specific stereochemistry (as compounds which could be used as effective drugs needed to have specific stereochemistry). He also developed a method whereby a molecule could be forced into the required configuration. Classic examples of this are the synthesis of reserpine and strychnine.

Woodward was one of the few chemists who successfully straddled the two disparate worlds of academia and business. He was closely associated with pharmaceutical industries such as Eli Lilly and Company, Merck & Co., Inc., Mallinckrodt Pharmaceuticals, Monsanto Company, Polaroid Corporation, and Pfizer. In 1963, the Woodward Research Institute was set up in Basel, Switzerland, by the Swiss pharmaceutical firm Ciba (Novartis International AG at present). He was simultaneously its director and the Donner Professor of Science at Harvard.

Honours

Woodward was a highly decorated organic chemist. He was honoured by many countries with their highest scientific medal or award. Among the important ones are: John Scott Medal of the Franklin Institute 1945, Baekeland Medal of the American Chemical Society 1955, Davy Medal of the Royal Society, London 1959, Roger Adams Medal of the American Chemical Society 1961, Pius XI Gold Medal of the Pontifical Academy of Sciences 1969, The US National Medal of Science in 1964, Nobel Prize in Chemistry 1965, Willard Gibbs Medal of the American Chemical Society 1967, Lavoisier Medal of the Société Chimique de France 1968, The Order of the Rising Sun, Second Class from the Emperor of Japan 1970, Hanbury Memorial Medal of The Pharmaceutical Society of Great Britain 1970, Pierre Bruylants Medal of the University of Louvain 1970, AMA Scientific Achievement Award 1971, Cope Award of the American

Chemical Society (shared with Roald Hoffmann), Copley Medal of the Royal Society, London, 1978. Woodward was conferred honoris causa degrees by more than twenty universities.

Lecturing Woodward style

Woodward was famous for his lectures, particularly his Thursday seminars. His Thursday seminars can be compared to Faraday's Friday evening lectures at the Royal Institution with one fundamental difference–Woodward never lectured for just an hour! As for his research, he was a meticulous planner of his seminars. It was the same routine every Thursday night in Converse Laboratory at Harvard. On arriving at the lecture hall, Woodward would neatly spread two large white handkerchiefs on the table. On one of them he would neatly arrange four or five rows of new coloured chalks and on the other handkerchief, cigarettes were arranged in neat rows. His students recall that he did not use a lighter after the first one was lit as he used the previous one to light the next one! For a number of years, he did not believe in using a slide projector for his lectures. When he finally decided to use slides, he made sure that he had a well trained projectionist who knew what Woodward wanted and also a spare bulb for the projector just in case the bulb in the projector fused! He preferred presenting intricate chemistry by drawing on the blackboard with multicolored chalks. He would start at the top left hand corner of the blackboard and work his way until he arrived at the last step at the bottom right corner. When he finished, the completed colourful diagram of the structure on the blackboard looked like beautiful art.

It was jokingly said that the unit of time 'the Woodward' was the time his longest lecture lasted. His other lectures were measured against this unit. His Thursday seminars at Harvard often lasted till the early hours of Friday. This marathon lectures were not just to impress his audience. He felt that if the subject required three hours or more to explain the steps of the synthesis, he was prepared to spend that much time lecturing. By the end of the seminar, the lecture hall would be filled with smoke and dust and the enthralled audience of students and colleagues would troop out bleary eyed.

Artist-chemist

Woodward was fascinated by every aspect of chemistry since his childhood. It was as if there was a bond between him and chemistry which strengthened his innate imagination and creativity. He confessed later that he was attracted by the sensuous aspect of chemistry- its change of colours, sounds of sloshing liquids that were common during the course of an experiment. He said in one of his lectures that *"chemistry would not exist for me without these physical, visual, tangible, sensuous things"* (From Benfey & Morris [4]). He also said that he chose a lifetime in chemistry instead of mathematics *"owing to preference to working in a realm where there was a physical restraint on fantas*y - and that *'one's ideas had to be applicable to the one physical environment we have"* (From Bowden & Benfey [7]). Woodward thought of chemistry 24x7, 365 days of the year. He often remarked that *"thinking of chemistry was a more exacting and demanding task than doing it"* (From Benfey & Morris [4]).

Woodward exhorted his students to *"write a formula in as many ways as possible. Each way may suggest different possibilities"* (From Benfey & Morris [4]). The many drawings he did before approaching the problem at hand having notations, exclamation marks all over revealed his thought processes. The elegant drawings containing precise progression of steps of reactions are the best tribute to his unique gift for visualising the complexities of the structures of the entire molecule. Even without any ability to decipher the profound scientific information contained in them, a lay person could enjoy the beauty of the drawings. And they were not computer generated graphics!

Woodward the man

Scheinbaum, one of Woodward's students said *"to be in the presence of Woodward was like young priests with the opportunity of being in the papal audience. There was an aura of greatness (about him) as he blessed us with a sense of his elegance and showed us the Sistine Chapels of chemical architecture"* (From Benfey & Morris [4]).

Woodward was worshipped by his students. They all acknowledged that working with him changed them in many ways. Derek Barton went so far as to say Woodward taught organic chemists of his time how to think.

Woodward had a towering personality. He was an extremely hard working scientist and expected the same from his students and post doctoral workers. When David Dolphin, a collaborator, joined Woodward, the first thing he wanted to know from Woodward was about vacations. David Dolphin recalled later Woodward's reply to the query was 'Well, I take Christmas Day off' (reminiscent of Faraday?). It is said that he required little sleep. He hated exercising and was passionate about blue colour. He was a voracious reader and read a variety of books which included detective novels and books on history. He was generous in acknowledging the contributions of his young students and collaborators. Contrary to his persona in the laboratory where he was a hard taskmaster, he was fond of playing practical jokes on his friends.

A practical joke

To celebrate the inauguration of the Conant Laboratories at Harvard University, a high profile dinner was hosted by the chairman's wife. Woodward knew that the hostess took pride in meticulously planning her dinners down to the smallest detail. Just before the guests were to be seated, Woodward apparently walked up to her and enquired what was the main dish at the dinner. When the hostess replied 'filet mignon' (beef steak), Woodward asked her if she also had fish on the menu as some guests might not eat meat on a Friday. The hostess took the bait and admitted that she had not thought of it and exclaimed "Oh Robert what shall I do now?", Woodward apparently said "Don't worry. It is only Tuesday today" and walked away calmly before the furious hostess could recover her poise.

Robert Burns Woodward passed away on July 8, 1979 of heart attack. His close friend and colleague at Harvard, Westheimer, remarked "He seemed immortal, we did not think he would ever die" (From Wallace & Gruber [8]).

References

1. Erin Elizabeth Milner, *The Grandfather of Organic Chemistry: Robert Burns Woodward, PhD,* Walter Reed Army Institute of Research, Silver Spring, MD.,doi: 10.1309/LM7LBJZCC20JLKSD (2010) LabMedicine, 41, 245-246. (ASCP).

2. From the interview of James F. Norris at RBW's graduation by the Boston Globe, 8 June 1937. (From Erika A. Crane, 'Life and Achievements of Robert Burns Woodward', Long Literature Seminar, July 13, 2009).

3. "Art and Science in the Synthesis of Organic Compounds: Retrospect and Prospect" in Pointers and Pathways in Research (Bombay: CIBA of India, 1963) (From the book by Mary Ellen Bowden, Otto Theodor Benfey, *'Robert Burns Woodward and the art of organic synthesis'*, A Publication of the Beckman Center for the History of Chemistry, 1992).

4. Otto Theodor Benfey, Peter John Turnbull Morris, (Eds)., *Robert Burns Woodward: Architect and Artist in the World of Molecules*, Chemical Heritage Foundation, 2001.

5. R. B. Woodward, *Experientia 12: supplement 2 (1955), (213-228).*

6. Derek H. R. Barton, *"Ingold, Robinson Winstein Woodward and I"* Bull. Hist. Chem. 19 (1996), pp (43-47).

7. (From an address Woodward delivered on receiving the American Chemical Society's Arthur C. Cope Award in Organic chemistry, 28 August, 1973) Mary Ellen Bowden, Otto Theodor Benfey, *'Robert Burns Woodward and the art of organic synthesis'*, A Publication of the Beckman Center for the History of Chemistry, 1992.

8. Doris B. Wallace, Howard E. Gruber, (Ed)., *Creative People at Work: Twelve Cognitive Case Studies,* Oxford University Press, 1989.

Robert B. Woodward - Biographical. Nobelprize.org. Nobel Media AB 2014. Web. 18 May 2015. <http://www.nobelprize.org/nobel_prizes/chemistry/laureates/1965/woodward-bio.html> (From *Nobel Lectures, Chemistry 1963-1970*, Elsevier Publishing Company, Amsterdam, 1972).

Organic Nobels

Organic chemistry is probably the area of chemistry with the largest number of practitioners and there have been many outstanding organic chemists. Although we have presented the academic biographies of three organic chemists from the early period (Wohler, Kekule and Fischer) and four from the modern period (Willstatter, Robinson, Woodward and Sanger), we should note that there have been a large number of Nobel laureates in the subject.

Before Woodward (Nobel prize, 1965), the following organic chemists received the Nobel prize in chemistry: 1912, V. Grignard (reagent) and P. Sabatier (hydrogenation); 1927, H. Wieland (bile acids); 1928, A. Windaus (sterols); 1930, H. Fischer (haemin); 1937, W. N. Haworth (carbohydrates) and P. Karrer (carotenoids and Vitamin A); 1938, R Kuhn (carotenoids and vitamins); 1939, A. F. J. Butenandt (sex hormones) and L. Ruzicka (polymethylenes and terpenes); 1950, O. Diels and K. Alder (diene synthesis); 1955, V.du Vigneaud (polypeptide hormone); 1957, A. Todd (nucleotides); 1963, K. Ziegler and G. Natta (high polymers).

After 1965, the following organic chemists have received the Nobel Prize: 1969, D. H. R. Barton and O. Hassel (conformation); 1979, H. C. Brown and G. Wittig (boron and phosphorus reagents for organic synthesis); 1984, R. B. Merrifield (chemical synthesis on a solid matrix); 1987, D. J. Cram, J-M. Lehn and C. J. Pedersen (molecules with structure-specific interactions of high selectivity); 1990, E. J. Corey (organic synthesis); 2000, A. J. Heeger, A. G. MacDiarmid and H. Shirakawa (conductive polymers); 2001, W. S. Knowles, R. Noyori and K. B. Sharpless (asymmetric synthesis); 2005, Y. Chauvin, R. H. Grubbs and R. R. Schrock (Metathesis in organic synthesis); 2010, R. F. Heck, E. Negishi and A. Suzuki (Pd-catalyzed cross couplings in organic synthesis).

21. FREDERICK SANGER (1918–2013)

A modest man with two Nobel prizes in chemistry

(From Nobelprize.org)

"*Sanger spent his career studying the three fundamental polymers of life – proteins, RNA and DNA. .... are made from long sequences of building blocks. It had long been known that DNA and RNA were made up of strings of just four bases, while proteins are more complicated, consisting of strings of 20 building blocks called amino acids. However, just knowing this is like understanding that sentences are made of letters but having no idea what order the letters come in. Sanger strove to decipher the order of DNA and RNA's bases and protein's amino acids*" (From Lorch [1]).

"*Frederick Sanger laid the bedrock on which some of the greatest achievements of 21st century science such as the Human Genome Project and all that has followed were built*" (From Lorch [1]).

"*Fred (Frederick Sanger) can fairly be called the father of the genomic era. His work laid the foundations of humanity's ability to read and understand the genetic code, which has revolutionized biology and is today contributing to transformative improvements in healthcare*" (From Farrar [2]).

Early years

Frederick Sanger was born on 13 August 1918, in a small village in Gloucestershire, England to Dr. Frederick Sanger Sr., a medical practitioner and Cicely Crewdson who came from a well to do family. Frederick Sanger Sr. went to China to work as a medical missionary but had to return to England earlier than intended due to ill health. He changed his religion and became a Quaker after the children were born. The children were raised as Quakers. Young Frederick Sanger was deeply influenced by the pacifist philosophy of Quakerism. Frederick and his elder brother Theodore were both home-schooled as the family was wealthy enough to employ a governess to take care of their early education. Frederick started his formal schooling when he was nine years old when he joined the Downs School, a preparatory school. He then joined Bryanston School in Dorset which followed a more liberal approach to education. The friendly atmosphere and dedication of teachers suited Frederick's temperament. The school was run by Quakers and influenced by Quakerism Frederick remained a conscientious objector. Academically, he was influenced by his father and elder brother from childhood and under their influence, he *"soon became interested in biology and developed a respect for the importance of science and the scientific method"* (From Nobelprize.org [3]).

Sanger was expected to follow in his father's footsteps and study medicine at the university but soon realized that he was temperamentally better suited to a career in science in which he could concentrate his efforts more on a single goal than in the medical profession (From Nobelprize.org [3]). This momentous decision resulted years later in two Nobel Prizes in chemistry!

Cambridge years

On securing good grades in his school certificate exam, he joined St John's College, Cambridge in 1936 as an undergraduate in science. He was particularly excited by the work being done in the biochemistry department as he felt that *"here was a way to really understand living matter and to develop a more scientific basis to many medical problems"* (From Nobelprize.org [3]).

Sanger opted for physics, chemistry, biochemistry and mathematics courses for Part I of his Tripos and biochemistry for part II. Tragedy struck Sanger while he was an undergraduate when he lost his parents, both succumbing to cancer. As an undergraduate, Sanger who believed in pacifism became a member of the Peace Pledge Union and whole heartedly participated in the activities of the Cambridge Scientists' Anti-War Group. He formally registered himself as a conscientious objector and was exempted from active war duties. His association with this group had an unanticipated happy consequence - he met Joan Howe whom he married after graduating in 1940. On completing his B A Tripos in 1939, Sanger spent an extra year in the Department of Biochemistry honing his skills in the subject. He later admitted that this was beneficial in his doctoral work. He registered for Ph. D in 1940 initially under N.W. "Bill" Pirie, who wanted Sanger to investigate whether edible proteins could be produced from grass and left him with *"a large bucket of frozen grass extract to be used for the investigation and left for a new job before it thawed"* (From Sanger [4]).

This proved to be a blessing in disguise as Sanger had to change his guide and the topic of his doctoral work. His new guide, Albert Neuberger, a post doctoral fellow, asked him to investigate the amino acid lysine's metabolism in animals. He found Neuberger inspiring both as research guide and a human being . Sanger enjoyed his association with Neuberger who gave him two valuable lessons – *"how to do research, both technically and as a way of life"* (From Nobelprize.org [3]), He successfully completed his Ph.D in 1943 with a thesis on "The metabolism of the amino acid lysine in the animal body". When Neuberger moved to National Institute for Medical Research in London soon after Sanger finished Ph.D in 1943, he decided to stay on in Cambridge and joined Charles Chibnall's group in the biochemistry department. This again proved to be a game changing decision.

Chibnall, a protein chemist, was working on bovine insulin. *"This was an especially exciting time in protein chemistry...and there seemed to be a real possibility of determining the exact chemical structure of these fundamental components of living matter"* (From Nobelprize.org [3]). Chibnall suggested that Sanger work on the amino groups in bovine insulin

– a protein. Sanger was excited by this suggestion as he felt that by using newly developed techniques of fractionation, separation and purification methods and chromatography, it might be possible to arrive at the actual chemical structure of amino acids. Chibnall encouraged Sanger as he was also convinced that time was ripe to unravel the chemical structure of insulin. However, most contemporary scientists felt the idea was controversial as the general belief was that the amino acids present in proteins were randomly arranged. This belief was so strong that the Medical Research Council refused to give a grant to pursue this research.

Insulin and thereafter

At the time Sanger started his work on the structure of insulin, scientists knew that proteins were chains of 20 different amino acids strung together like a chain of coloured beads but had no inkling on how or in what sequence these were strung together. It was also believed that proteins were amorphous substances. Sanger decided to study the structure of bovine insulin by analyzing the amino acid sequence because insulin was a protein with a short chain of 51 amino acids. It was readily available in pure form and could be bought across the counter from a chemist's shop. Also, it was important for health care as it was the standard drug to treat diabetes. Sanger developed a new method which he called the jigsaw puzzle method for sequencing the amino acid sequence in insulin. This method involved cutting the insulin chain into smaller pieces, marking the end amino acid and splicing it off from the insulin chain. By repeating this process, Sanger was able to arrive at the conclusion that there were two peptide chains A and B in a molecule of insulin and these two were linked to each other by disulphide bonds. Sanger succeeded in determining the amino acid sequence of polypeptide chain B in 1951 as the chain was shorter and less complex, and followed this by sequencing chain A in 1952. Based on this experimental evidence, Sanger established that proteins were chemical substances with definite and unique compositions. He also concluded that the polypeptide chains of the protein insulin had precise sequences of amino acids and therefore every protein had its own unique sequence. Even though the sequence of amino acids in insulin was fairly simple and straight forward, it took Sanger twelve long years of concentrated and dogged work to arrive at this historic conclusion and this understanding paved the way for unraveling the DNA code.

Jigsaw puzzle method

Sanger reacted fluorodinitrobenzene (FDNB), later called the "Sanger Reagent", with the exposed amino groups in the protein and followed this up by hydrolysing the chain into shorter peptides using either hydrochloric acid or the enzyme trypsin. This peptide mixture was fractionated on a sheet of filter paper by chromatography. The peptide fragments of insulin moved to different positions on the paper creating a distinct pattern or "fingerprints".

Fluorodinitrobenzene gave a distinct yellow colour to the peptide chain from the N-terminus and the amino acid at the end of the peptide could be identified by finding out which dinitrophenyl-amino acid was present there on acid hydrolysis. Sanger repeated this procedure and succeeded in determining the sequences of a number of peptides. He then reassembled the peptide pieces into longer sequences and solved the complete structure of insulin. After twelve years of conducting many experiments, Sanger announced in 1951 that proteins had a definite chemical composition and that there was a unique sequence in every protein. The final piece of the jigsaw puzzle fitted into the slot when Sanger and coworkers solved the role of the three linking disulfide bonds –two interchain linking A and B chains and one intrachain on chain A in 1955.

The period that followed was in stark contrast to the nearly twelve years of intense work on insulin. Sanger's response to this period of "*lean years with no major success*" is best captured in this quote "*I think these periods occur in most people's research careers and can be depressing and sometimes lead to disillusion. I have found the best antidote is to keep looking ahead. When an experiment is a complete failure it is best not to spend too much time worrying about it but rather get on with planning and becoming involved in the next one. This is always exciting and you soon forget your troubles*" (From Lorch [1]).

Sanger true to his belief moved on and got involved in planning the next phase of his research. He was proved right as during this *"lean years with no major success"* he was awarded the 1958 chemistry Nobel Prize *"for his work on the structure of proteins, especially that of insulin"* (From Nobelprize.org [5]).

Life after the first Nobel Prize

"This award had an important and stimulating effect on my subsequent career. I had remained in Cambridge concentrating only on basic research and avoiding as far as possible teaching or administrative responsibilities. This recognition of my work gave me renewed confidence and enthusiasm to continue in this way of life, which I enjoyed. It also enabled me to obtain better research facilities and, even more important, to attract excellent colleagues" (From Nobelprize.org [3]).

The Nobel Prize vindicated Sanger's way of doing research. He worked on his own devising and conducting experiments in the laboratory without a big research group. He did not teach nor was involved in administrative work. As he himself remarked *"Of the three main activities involved in scientific research, thinking, talking and doing, I much prefer the last and am probably best at it. I am all right at the thinking, but not much good at the talking"* (From Lorch [1]).

"Sanger was the Paul Dirac of chemistry; technically brilliant, wedded to his work, reserved to a fault, averse to teaching and administration, a scientist's scientist in every way" (From Jogalekar [6]).

Sequencing DNA

The most exciting period of Sanger's research started in 1962 when he moved to the newly established Laboratory of Molecular Biology set up by the Medical Research Council. Here, research was focussed on understanding genes in general and DNA and RNA in particular. Scientists were aware for a long time that both DNA and RNA strings with just four bases were simpler than protein strings which had 20 amino acids. But this minimal knowledge gave no clue to the sequence of the order of bases in DNA and RNA and the amino acids in proteins. Sanger got interested in the hunt for the code of RNA and DNA.

"Although at the time it seemed to be a major change from proteins to nucleic acids, the concern with the basic problem of "sequencing" remained the same." Sequencing was the focus of his research since 1943, *"both because of its intrinsic fascination and my conviction that a knowledge of sequences could contribute much to our understanding of living matter"* (From Nobelprize.org [3]). The major problem Sanger faced was how to get a pure RNA fragment to sequence. Sanger and his co-workers Bart Barrell, Alan Coulson and George Brownlee spent the 60s and 70s trying to develop a reliable method to sequence DNA and RNA. By 1971 scientists had only succeeded in sequencing a small string of DNA of 12 bases. He began work on developing newer methods and as an unexpected fallout, Sanger and his coworker Kjeld Marcker discovered in 1964, formylmethionine tRNA which is responsible for protein synthesis in bacteria. This work to sequence RNA was being aggressively pursued by Robert Holley's group at Cornell University and in 1965 they succeeded in sequencing the tRNA molecule with 77 nucleotides. Two years later, Sanger's group succeeded in determining the sequence of 5S ribosomal RNA's 120 nucleotides (from *Escherichia coli*).

After his success in sequencing RNA's nucleotides, Sanger shifted his focus to sequencing DNA nucleotides or sequencing genes. He discovered that this work needed an altogether different approach. He had to devise ways to use DNA polymerase I from *E. coli* and copy single stranded DNA. After years of intense work, Sanger and Alan Coulson published in 1975 a new sequencing technique in which they used DNA polymerase (an enzyme) with radiolabelled (labelled with radio active atoms) nucleotides to copy DNA fragments. These nucleotides unlike the ones in DNA molecule terminate DNA strand to grow further or act as 'chain terminators'. Sanger called this technique "Plus and Minus" technique (From Nobelprize.org [7]). Even though this technique succeeded in sequencing close to 80 nucleotides at a time and was a vast improvement over the then available techniques, it was very time-consuming and labour intensive to sequence large DNA based genomes. Sanger and his group succeeded in sequencing the first DNA based genome, the single-stranded bacteriophage φX174. They were surprised to find the overlapping of the coding regions of some genes in bacteriophage φX174.

During this period, Sanger developed many novel techniques for sequencing DNA bases and the technique of "reading" DNA, such as the use of special bases called chain terminators for more efficient method of cloning to replicate both the strands of DNA. Many research groups in other countries, especially in the US were also working on this problem. Sanger's group succeeded in sequencing almost all the nucleotides (5,386) of single-stranded bacteriophage φX174. This was a great achievement as this was the first DNA-based genome to be sequenced almost completely. Around the mid 70s, a breakthrough was achieved by the American team led by Gilbert and the UK group led by Sanger. While Gilbert and Maxam followed the 'chemical cleavage protocol', Sanger used the 'dideoxy' sequencing method or the 'Sanger method' which Sanger himself called *"DNA sequencing with chain-terminating inhibitors"*. Sanger's method became the universally accepted method for DNA replication and decoding as it imitated Nature's way of replicating DNA. It was faster and more practical. This technique made it possible to sequence quickly and accurately long stretches of DNA. Sanger and his colleagues used this method to sequence human mitochondrial DNA with 16,569 base pairs. This was a major breakthrough as this DNA is present in all cells present in the human body.

Sanger and co workers announced the successful sequencing of bacteriophage λ in a paper published in Nature. *"A DNA sequence for the genome of bacteriophage φX174 of approximately 5,375 nucleotides has been determined using the rapid and simple 'plus and minus' method. The sequence identifies many of the features responsible for the production of the proteins of the nine known genes of the organism, including initiation and termination sites for the proteins and RNAs. Two pairs of genes are coded by the same region of DNA using different reading frames"* (From Sanger et al. [8]).

Sanger received his second Nobel Prize in chemistry in 1980 along with Walter Gilbert of Harvard University. They shared half of the chemistry Nobel Prize "*for their contributions concerning the determination of base sequences in nucleic acids*" and Paul Berg of Stanford University was awarded the other half of the prize "*for his fundamental studies of the biochemistry of nucleic acids, with particular regard to recombinant DNA*" (From Nobelprize.org [9]).

> Apart from Marie Curie (Physics, 1903 and Chemistry, 1911), Linus Pauling (Chemistry, 1954 and Peace, 1962) and John Bardeen (Physics, 1956 and 1972), Fred Sanger is the only other scientist to receive two Nobel prizes. He also had the unique distinction of being the only person to receive two prizes in chemistry.

From sequencing genomes of 12 bases to the Human Genome Project

In 1977, Sanger and colleagues designed a path breaking method for sequencing DNA. The unique feature of this method was that it was similar to the method used by DNA for replication. Here dideoxynucleotides are used to terminate DNA replication. Dideoxynucleotides are similar to the normal nucleotides found in the DNA strands except that while there is a hydroxyl group at 3' carbon position in natural nucleotides, there is a hydrogen at that place in dideoxynucleotides. When a dideoxynucleotide is introduced into a growing DNA strand, the growth of the chain stops as a bond between the dideoxynucleotide and the next nucleotide cannot be formed as dideoxynucleotide lacks the OH group essential for the chain to continue. This method is named 'dideoxy' method after the critical role played by dideoxynucleotides. It is also called the chain-termination method for DNA sequencing. It is popularly known as the "Sanger method" acknowledging the genius of Sanger. By using this method, it became possible to sequence long stretches of DNA accurately and speedily. Further, this method made the ambitious dream of sequencing the entire human genome become a reality via the human genome project. Now genomes with more than 3,000,000,000 base-pairs can be studied using the Sanger method.

Impact of Sanger's work

"Sanger's work unlocked the chemical secrets that underlie genes - the basic building blocks of life - and laid the foundation for genetic engineering and the Human Genome Project, a unique effort to spell out the chemical structure of every gene in the human body" (From The Telegraph [10]).

"*In the 1950s when the very identity of proteins was unclear, Sanger came up with a simple chemical method for identifying individual amino acids in a protein or peptide by tagging them with a brightly colored chemical handle. The first protein whose sequence he unravelled - insulin - was both a revelation and a technical tour de force. Combined with Linus Pauling's work on delineating the basic structure of proteins, Sanger essentially set protein science on a rational basis*" (From Jogalekar [6]).

Honours

Apart from the unique distinction of being the only chemist to receive two Nobel prizes for chemistry (1958, 1980) Sanger received innumerable medals and honours starting from Corday–Morgan Medal in 1951. He was elected as a Fellow of the Royal Society in 1954, Commander of the Order of the British Empire (1963), Royal Medal (1969), Gairdner Foundation International Award (1971), William Bate Hardy Prize (1976), Copley Medal (1977), G. W. Wheland Award (1978), Louisa Gross Horwitz Prize of Columbia University and Albert Lasker Award for Basic Medical Research (1979), Order of the Companions of Honour (1981), Corresponding Fellow of the Australian Academy of Science (1982), Order of Merit (1986).

What makes Sanger unique?

Sanger was a *"courteous, serious-minded man of strong socialist opinions", his "thin bespectacled figure, habitually dressed in academic-casual v-necked sweater, open-necked shirt and rubber-soled shoes, was a familiar sight in Cambridge for many years"* (From

The Telegraph [10]). He cycled to the lab everyday in rain or sunshine. Even though Sanger was reclusive by nature and could go unnoticed even in a crowd of scientists, he was extremely friendly and helpful to the young people working in his laboratory and inspired them by example rather than by words of wisdom. He could be in the midst of a crowd of distinguished scientists yet be mistaken for a gardener of a Cambridge college. But make no mistake, *"he knew that he was an extraordinary scientist, and when the occasion demanded it he was prepared to say so. When colleagues assembled after the announcement of his second Nobel Prize, one praised his characteristic modesty. Sanger responded: 'I want you all to know that I think that I am bloody good'"* (From Walker [11]).

> One of the authors of this book (CNR Rao) was in a Christmas party in Cambridge in 1983 where Sanger was also present. Sanger was standing by the window wearing his old woolen jacket. Someone came and asked who is that man standing by the window? He thought it would be an intruder or one of the lesser beings in the college. He was told that the man near the window was a scientist with two Nobels.

Sanger's attitude towards honours and awards that he received reminds one of Faraday. According to Sanger: *"You get a nice gold medal, which is in the bank,"* he explained. *"And you get a certificate, which is in the loft. I could put it on the wall, I suppose. I was lucky and happy to get it, but I'm more proud of the research I did. There are some people, you know, who are in science just to get prizes. But that's not what motivates me"* (From The Telegraph [10]). Accordingly plaques, certificates or citations were not on the walls of his house. Fred Sanger according to his friends and colleagues was the most self-effacing person you could hope to meet. His modesty was not a put on. He said *"I was just a chap who messed about in his lab"* (From "Frederick Sanger." Wikipedia [12]) *and "academically not brilliant"* (From Lorch [1]). Like Faraday before him, he refused knighthood because he preferred to be just Fred to his colleagues and friends and did not want to be addressed as 'Sir Fred Sanger'. *"A knighthood makes you different, doesn't it, and I don't want to be different"* (From The Telegraph [10]).

Personal life

"I was married to Margaret Joan Howe in 1940. Although not a scientist herself she has contributed more to my work than anyone else by providing a peaceful and happy home" (From Nobelprize.org [3]).

Margaret Joan Sanger shared Sanger's anti war sentiments and pacifist philosophy. Much like Eva Helen Pauling, she supported her husband's preoccupation with his scientific career but unlike Eva Helen Pauling, she did not participate in high profile public protests. Sanger acknowledged her invaluable help in his autobiography. They had three children – two sons and a daughter. For a man who confessed he did not like talking, Sanger had quick wit and a gift of repartee. When Joanna Rose, science writer asked him how he felt about getting the second Nobel Prize, he said *"It's much more difficult to get the first prize than to get the second one"*, he said, *"because if you've already got a prize, then you can get facilities for work, and you can get collaborators, and everything is much easier"* (From Nobelprize.org [13]). Again, when one of the awestruck students of his old school interviewed him and asked what else was left for him to do after winning two Nobel Prizes, Sanger was said to have remarked *"Win a third Nobel prize"* (From Chemistryworldblog [14]). His reply to the request of the founding Director of the Wellcome Institute, John Sulston (himself a Nobel Laureate) for permission to rename the institute The Sanger Institute, Sanger agreed but with a rider that *"It had better be good"* (From Sanger Institute [15]).

Last years with roses

In spite of all the accolades, when he reached 65, the formal age of retirement, Sanger surprised everybody (and probably himself) when he decided to make a drastic change in his lifestyle by completely giving up research. He was a passionate gardener but had not found time to enjoy his hobby. He decided to heed the call of his rose bushes and spent the rest of his life tending to his roses and 'messing about in boats' and never went back to the lab!

Sanger on his decision to leave the lab at the age of 65

"I had not thought about retirement until I suddenly realized that in a few years I would be 65 and would be entitled to stop work and do some of the things I had always wanted to do and had never had time for. The possibility seemed surprisingly attractive, especially as our work had reached a climax with the DNA sequencing method and I rather felt that to continue would be something of an anticlimax. I have greatly enjoyed the new life-style, but also because the aging process was not improving my performance in the laboratory and I think that if I had gone on working I would have found it frustrating and have felt guilty at occupying space that could have been available to a younger person. For more than 40 years I had had wonderful opportunities for research, and had been given the chance to fulfill some of my wildest dreams" (From Sanger [4]).

The Sanger Centre was officially inaugurated by Sanger on the 4th October, 1993 with less than 50 people working there. It now is one of the leading centres and hub of Human Genome Project. For the modest Sanger, the climax of his career was his DNA sequencing method and was of greater value than the Nobel Prize he got for his work.

Sanger died peacefully in his sleep at the age of 95 in his beloved Cambridge. Echoing the admiration of his colleagues, fellow biologists and scientists at large, Sanger Institute in its announcement of his death stated *"Sadly Fred died on Tuesday 19 November 2013 at the age of 95.* ***If anyone was the father of genomics it was this quiet, determined, modest man of strong opinions****"* (From Sanger Institute [15]).

References

1. Mark Lorch, '*Frederick Sanger's achievements cannot be overstated*', University of Hull, Article originally published in The Conversation.com, 22 November 2013.
2. Jeremy Farrar, *Frederick Sanger: 1918-2013,* Wellcome Trust, 20 November 2013.

3. Frederick Sanger - Biographical. *(From Les Prix Nobel. The Nobel Prizes 1980, Editor Wilhelm Odelberg, [Nobel Foundation], Stockholm, 1981) Nobelprize.org.* Nobel Media AB 2014. Web. 11 May 2015. <http://www.nobelprize.org/nobel_prizes/chemistry/laureates/1980/sanger-bio.html>
4. Frederick Sanger, *Sequences, Sequences, and Sequences*, Ann. Rev. Biochem. 1988, 57:1-28.
5. "The Nobel Prize in Chemistry 1958". *Nobelprize.org.* Nobel Media AB 2014. Web. 11 May 2015. <http://www.nobelprize.org/nobel_prizes/chemistry/laureates/1958/>
6. Ashutosh Jogalekar, *Winning two Nobel Prizes, turning down knighthoods: The legacy of Fred Sanger (1918-2013*), Scientific American.com, November 20, 2013.
7. "Frederick Sanger - Nobel Lecture: *Determination of Nucleotide Sequences in DNA". Nobelprize.org.* Nobel Media AB 2014. Web. 11 May 2015. <http://www.nobelprize.org/nobel_prizes/chemistry/laureates/1980/sanger-lecture.html>
8. *F.* Sanger et.al., *Nucleotide Sequence of Bacteriophage öX174 DNA,* Nature 265, 687-695, (1977).
9. "The Nobel Prize in Chemistry 1980". *Nobelprize.org.* Nobel Media AB 2014. Web. 12 May 2015. <http://www.nobelprize.org/nobel_prizes/chemistry/laureates/1980/>
10. Frederick Sanger, OM, Science Obituaries, The Telegraph, 20 November 2013.
11. John Walker, *Frederick Sanger (1918-2013)*, Nature 505, 27, 2 Jan 2014.
12. "Frederick Sanger." *Wikipedia, The Free Encyclopedia.* Wikipedia, The Free Encyclopedia, 5 May. 2015. Web. 13 May. 2015.
13. *Interview with Frederick Sanger (13 minutes),* An interview with Dr. Frederick Sanger by Joanna Rose, science writer, 9 December 2001. Video Player". *Nobelprize.org.* Nobel Media AB 2014. Web. 13 May 2015. <http://www.nobelprize.org/mediaplayer/index.php?id=360>
14. *My hero: Frederick Sanger*, Posted by Bea on Thu 31 Mar 2011 Chemistryworldblog, RSC.
15. *Frederick Sanger*, Wellcome Trust Sanger Institute.